Mathematical Methods for Geophysics and Space Physics

Mathematical Methods for Geophysics and Space Physics

William I. Newman

Princeton University Press

Princeton and Oxford

Library of Congress Cataloging-in-Publication Data

Names: Newman, William I., author.
Title: Mathematical methods for geophysics and space physics /
    William I. Newman.
Description: Princeton, New Jersey ; Oxford : Princeton University
    Press, [2016] | ©2016 | Includes bibliographical references and index.
Identifiers: LCCN 2015040966 | ISBN 9780691170602 (hardback ;
    alk. paper) | ISBN 0691170606 (hardback ; alk. paper)
Subjects: LCSH: Geophysics-Mathematics. | Cosmic physics-Mathematics.
Classification: LCC QC809.M37 N49 2016 | DDC 520.1/51-dc23
    LC record available at http://lccn.loc.gov/2015040966

British Library Cataloguing-in-Publication Data is available

This book has been composed in Lucida Bright and typeset
by T&T Productions Ltd, London

Printed on acid-free paper ∞

Printed in the United States of America

10 9 8 7 6 5 4 3 2 1

Contents

Contents

Preface

Graduate students in the earth sciences, particularly those in geophysics and atmospheric, oceanographic, planetary, and space physics, as well as astronomy, require a substantial degree of mathematical preparation—for the sake of brevity, we will simply refer to these application areas as being in geophysics. While there is significant overlap between their needs and those of graduate students in physics or in applied mathematics, there are important differences in the preparation needed and, notably, the sequence of presentation required as well as the overall quantity of material that is necessary. Most textbooks that address mathematical methods for physics and engineering begin from the standpoint that the student already knows the underlying equations, generally partial differential equations, but needs to learn how to solve them. Since the background of most entering or second-year graduate students in geophysics is highly variable, I felt it necessary to provide derivations in a number of circumstances for those equations to help students appreciate better where they arise and how their solution must be addressed. Moreover, most mathematical methods textbooks were published before the renaissance in thinking, especially about geophysical problems, that introduced the concepts of chaos and complexity, as well as the significance of probability and statistics and of numerical methods. Significant attention is given to the ordinary and partial differential equations that have played a pivotal role in the evolution of geophysics. In addition, in order to round out our treatment of mathematical methods, a succinct survey of statistical and computational issues is introduced. A brief but comprehensive summary of solution methods is presented, including many exercises. In so doing, it is my hope that this book will address that need during the course of one academic semester or quarter. In essence, we treat some central problem areas in depth, while providing a measure of literacy in others.

Students also can find helpful materials in the following works. The text that is closest to our presentation is that due to

Mathews and Walker (1970) which is out of print. A relatively contemporary text on the topic is that of Arfken and Weber (2005), but its newest edition (Arfken et al., 2013) has become a "comprehensive guide"; a helpful lead-in to the latter, designed more for advanced undergraduates, is Weber and Arfken (2004). The graduate textbook by Stone and Goldbart (2009) is also helpful, although the examples selected are drawn from physics and have a more formal flavor. Finally, the classic two-volume definitive texts on the subject are those by Morse and Feshbach (1999) and by Courant and Hilbert (1962). While the former is now back in print, the latter remains out of print. Regarding specific applications to geophysics and planetary, atmospheric, oceanographic, and space physics, chapters in existing graduate-level textbooks in those specialties contain appropriate derivations. As we encounter each new topic, additional citations to reference materials will be provided. We shall attempt to integrate some of the most important of these into this book.

Given the time available in a single academic quarter or semester, we are fundamentally limited in the quantity of material that can be presented. Basically, we provide an overarching survey of the relevant issues, a brief treatment of how to treat these problems, and an indication for each of these topics where the student can find a more thorough and rigorous treatment. Our objective is to give each student sufficient instruction to solve elementary problems and then advance to more exhaustive treatments of the individual topics, whether they originated in geophysics and its associated disciplines, physics, astronomy, or engineering.

The first chapter reviews many mathematical preliminaries that students should have studied previously, but also serves as a review. Vectors, indicial or "Einstein" notation, vector operators, cylindrical and spherical geometry, and the theorems of Gauss, Green, and Stokes are presented here. Since the focus of this chapter is on geometry, we introduce matrices in the context of the rotation of vectors. Then, we present tensors, which are matrices whose physical properties remain unchanged under a rotation and preserve other physical (e.g., variational) principles, including a very brief description of the eigenvalue problem. Here we introduce the concept of generalized functions through the Dirac δ function, and some of its relatives, inasmuch as they will form the basis later for our treatment of Green's functions. We present a number of assignment problems.

In the second chapter, we review features of ordinary differential equations. The Laplacian operator in partial differential equations permits the use of the method of separation of variables, which yields a set of second-order ordinary differential equations in the different geometric variables. We introduce the concept of Green's functions. Accordingly, we treat the separation of variables issue from the standpoint of ordinary differential equations and we introduce the derivation underlying Bessel functions and spherical harmonics, including the Legendre polynomials. (We complete the discussion of Poisson's, Laplace's, and Helmholtz's equation in chapter 4 because of their utility in solving partial differential equations of elliptic type.) We introduce problems describable by coupled ordinary differential equations, which, ultimately, provide the basis for chaos theory and are largely overlooked in classical mathematical methods of physics textbooks. Geophysical examples provide a wonderful testbed for ordinary differential equation approaches. For example, efforts to model the geodynamo using the interaction of mechanical and electrical components yielded strictly cyclical behavior with no field reversals. Efforts to resolve this problem demonstrated an epiphanic paradigm shift in moving to systems with three equations, such as the Lorenz model for convection and turbulence. This chapter also provides hands-on experience in performing perturbation theory analysis. Since chaotic behavior often yields fractal geometry, as in the Lorenz model trajectory, we provide a brief survey of fractal concepts and applications, as well as mappings as an adjunct to understanding transition to chaos.

In the third chapter, we introduce the evaluation of integrals, including a brief overview of complex analysis and elementary contour integration, saddle point methods, and some special problems in geophysics that yield elliptic integrals. We continue to address integral transforms following a brief introduction to Fourier series and transforms. We prove the sampling theorem and describe the phenomenon of aliasing. While these latter topics are overlooked in most textbooks, they play an important role in geophysics, particularly in the context of data collection and analysis. We introduce the fast Fourier transform and some approximation methods for spectral analysis. We conclude this chapter by briefly touching upon Laplace transforms and the Bromwich integral, and we introduce some integral equations, including the Abel and Radon transforms, as well as the Herglotz–Wiechert problem of seismology.

In chapter 4, we introduce the fundamental partial differential equations of mathematical physics, in general, and geophysics, in particular. We whet the student's appetite by introducing the three fundamental types of partial differential equations that are pervasive in geophysics: the wave equation, the potential equation, and the diffusion equation. This chapter embeds practical examples of real-world problems with the theory. Classic mathematical methods of physics books rarely provide examples, especially those that are appropriate to the earth sciences. Remarkably, some of the most beautiful yet practical examples of these types of equations appear in geophysics. We introduce, for linear problems, integral transform methods, and introduce eigenfunctions, eigenvalues, and Green's functions in those time-dependent contexts. We exploit these methods to solve both the diffusion equation and the wave equation in three dimensions. We employ spherical harmonics, introduced in the second chapter, to solve the gravitational potential equation relating a planet's mass distribution to its potential in three dimensions. Further, we exploit Fourier methods in order to identify dispersion relations for linear problems, including the role of diffusion and dispersion. At this stage, we associate with dispersion relations for partial differential equations the role of instability. Perturbation theory in this context is presented via a simple example, the propagation of sound in a fluid. However, since partial differential equations incorporate an infinite number of modes—associated with spherical harmonics, for example—the chaotic nature of a fundamentally infinite degree of freedom system underscores what is called *complexity*. We consider collective, nonlinear modes of behavior as exemplified by solitary waves and, especially, solitons. As illustrations, we derive the solution for solitary waves exemplified by Burgers's equation and for solitons via the Korteweg–de Vries equation. Scaling arguments underlying the emergence of turbulence are presented, as well as a simple derivation for the Kolmogorov spectrum.

The remaining chapter surveys two topics that are central to modern geophysics yet have been orphaned from essentially all elementary treatments. We briefly survey topics in probability and statistics, including the binomial, Poisson, and Gaussian (normal) distributions as well as the central limit theorem. A sketch is provided for methods of random number generation, central to Monte Carlo simulation. We also identify some of the themes associated with regression and the fitting of experimental data. Finally, we survey some questions emergent from numerical methods. Here, we briefly address the nature of computational and round-off errors. As an example, we survey the determination of the roots of polynomials, which play a fundamental role in the dispersion relations of modern geophysics. We provide a brief overview of numerical methods of solving ordinary and partial differential equations, with a focus on finite difference methods, but mention spectral approaches.

As is evident, this textbook provides a whirlwind survey of many topics and helps bring together many different concepts yet provide a brief practical introduction to problem solving in geophysics. This book was developed in consultation with my colleagues and is the outcome of several offerings at UCLA of this survey course to entering and second-year graduate students in geophysics and planetary and space physics, but was also designed to be helpful to students in allied disciplines, including atmospheric and ocean sciences, and in physics and astronomy. We very much hope that this volume will help stimulate thinking about these problem areas and further investigation and study of the different topics reviewed.

While completing this volume, my editor asked me to provide a cover image for this book and recommended that a photograph be adopted instead of a geometrical design or blank cover as is so often employed in technical books. This presented a special challenge inasmuch as how could a photograph convey what underscores the mathematics implicit to the earth, planetary, and space sciences? What kind of image would capture the outcome of a combination of many different geologic events? Yellowstone National Park is a truly special place, and the Grand Canyon of the Yellowstone is a focal point for much of its varied geologic history. This area was shaped by a caldera eruption 600,000 years ago and a series of lava flows. The area was also faulted by the caldera dome before the eruption. The site of this canyon was possibly established by this faulting, which magnified the rate of erosion. Glaciation also took place, although glacial deposits are largely absent. This photograph features the Lower Falls, 308 feet in height, as viewed from Lookout Point. The rich colors of the rock in this photograph are likely an outcome of the hydrothermal alteration of the rhyolite containing different iron compounds and their subsequent "cooking." Exposure to the elements and oxidation added to this effect, and are not due to sulfur. The falling water provides a quick reminder of the power of the flow. Thinking about all of the various physical and chemical effects present in creating this scene, it is clear how this image captures so many different influences and that challenge of providing a quantitative description of them. I took this photograph on August 24, 2009, with a Sony A350 DSLR at F8 with a 1/320-second exposure time using a 160-mm zoom lens.

Acknowledgements

I gratefully acknowledge the advice and comments provided by the many students to whom I have had the privilege of teaching this material. Their comments and advice were invaluable. I have benefited as well from the advice of many of my colleagues, notably Paul Davis, Roger Grimshaw, Richard Lovelace, Darryl Holm, Didier Sornette, Mac Hyman, Aric Hagberg, Jim McWilliams, Jon Aurnou, Ron Powell, Werner Israel, Saul Teukolsky, and Michael Efroimsky, among others. I especially want to thank Nat Hamlin, Paul Roberts, Bruce Bills, Philip Sharp, Rick Schoenberg, Bernd Krauskopf, and Fritz Busse for detailed comments on the manuscript. Finally, I wish to thank my editor, Ingrid Gnerlich, for her steadfast support. I also wish to thank my copyeditor, Teresa Wilson, for making the final phase of working with the manuscript relatively painless and for her painstaking attention to detail.

William I. Newman
University of California, Los Angeles
June 16, 2015

Mathematical Methods for Geophysics and Space Physics

CHAPTER ONE

Mathematical Preliminaries

The underlying theory for geophysics, planetary physics, and space physics requires a solid understanding of many of the methods of mathematical physics as well as a set of specialized topics that are integral to the diverse array of real-world problems that we seek to understand. This chapter will review some essential mathematical concepts and notations that are commonly employed and will be exploited throughout this book. We will begin with a review of vector analysis focusing on indicial notation, including the Kronecker δ and Levi-Civita ϵ permutation symbol, and vector operators. Cylindrical and spherical geometry are ubiquitous in geophysics and space physics, as are the theorems of Gauss, Green, and Stokes. Accordingly, we will derive some of the essential vector analysis results in Cartesian geometry in these curvilinear coordinate systems. We will proceed to explore how vectors transform in space and the role of rotation and matrix representations, and then go on to introduce tensors, eigenvalues, and eigenvectors. The solution of the (linear) partial differential equations of mathematical physics is commonly used in geophysics, and we will present some materials here that we will exploit later in the development of Green's functions. In particular, we will close this chapter by introducing the ramp, Heaviside, and Dirac δ functions. As in all of our remaining chapters, we will provide a set of problems and cite references that present more detailed investigations of these topics.

1.1 Vectors, Indicial Notation, and Vector Operators

This book primarily will pursue the kinds of geophysical problems that emerge from scalar and vector quantities. While mention will be made of tensor operations, our primary focus will be upon vector problems in three dimensions that form the basis of geophysics. Scalars and vectors may be regarded as tensors of a specific rank. *Scalar* quantities, such as density and temperatures, are *zero-rank* or *zero-order* tensors. *Vector* quantities such as velocities have an associated direction as well as a magnitude. Vectors are *first-rank* tensors and are usually designated by boldface lower-case letters. *Second-rank tensors*, or simply *tensors*, such as the stress tensor are a special case of square matrices. Matrices are generally denoted by boldface, uppercase letters, while tensors are generally denoted by boldface, uppercase, sans-serif letters (Goldstein et al., 2002). For example, M would designate a matrix while T would designate a tensor. [There are other notations, e.g., Kusse and Westwig (2006), that employ overbars for vectors and double overbars for tensors.] Substantial simplification of notational issues emerges upon adopting *indicial* notation.

In lieu of x, y, and z in describing the Cartesian components for position, we will employ x_1, x_2, and x_3. Similarly, we will denote by $\hat{e}_1$, $\hat{e}_2$, and $\hat{e}_3$ the mutually orthogonal *unit vectors* that are in the direction of the x_1, x_2, and x_3 axes. (Historically, the use of e emerged in Germany where the letter "e" stood for the word *Einheit*, which translates as "unit.") The indicial notation implies that any repeated index is summed, generally from 1 through 3. This is the *Einstein summation convention*.

It is sufficient to denote a vector v, such as the velocity, by its three components (v_1, v_2, v_3). We note that v can be represented vectorially by its component terms, namely,

$$v = \sum_{i=1}^{3} v_i \hat{e}_i = v_i \hat{e}_i. \tag{1.1}$$

Suppose T is a tensor with components T_{ij}. Then,

$$T = \sum_{i=1,j=1}^{3} T_{ij} \hat{e}_i \hat{e}_j = T_{ij} \hat{e}_i \hat{e}_j. \tag{1.2}$$

We now introduce the *inner product*, also known as a scalar product or dot product, according to the convention

$$u \cdot v \equiv u_i v_i. \tag{1.3}$$

Moreover, we define u and v to be the lengths of u and v, respectively, according to

$$u \equiv \sqrt{u_i u_i} = |u|; \qquad v \equiv \sqrt{v_i v_i} = |v|; \tag{1.4}$$

we can identify an angle θ between $\boldsymbol{u}$ and $\boldsymbol{v}$ that we define according to

$$\boldsymbol{u} \cdot \boldsymbol{v} \equiv uv \cos \theta, \qquad (1.5)$$

which corresponds directly to our geometric intuition.

We now introduce the Kronecker δ according to

$$\delta_{ij} = \begin{cases} 1 & \text{if } i = j, \\ 0 & \text{if } i \neq j. \end{cases} \qquad (1.6)$$

The Kronecker δ is the indicial realization of the identity matrix. It follows, then, that

$$\hat{\boldsymbol{e}}_i \cdot \hat{\boldsymbol{e}}_j = \delta_{ij}, \qquad (1.7)$$

and that

$$\delta_{ii} = 3. \qquad (1.8)$$

This is equivalent to saying that the *trace*, that is, the sum of the diagonal elements, of the identity matrix is 3. An important consequence of Eq. (1.7) is that

$$\delta_{ij}\hat{\boldsymbol{e}}_j = \hat{\boldsymbol{e}}_i. \qquad (1.9)$$

A special example of these results is that we can now derive the general scalar product relation (1.3), namely,

$$\boldsymbol{u} \cdot \boldsymbol{v} = u_i\hat{\boldsymbol{e}}_i \cdot v_j\hat{\boldsymbol{e}}_j = u_iv_j\hat{\boldsymbol{e}}_i \cdot \hat{\boldsymbol{e}}_j = u_iv_j\delta_{ij} = u_iv_i, \qquad (1.10)$$

by applying Eq. (1.7).

We introduce the Levi-Civita or permutation symbol ϵ_{ijk} in order to address the *vector product* or *cross product*. In particular, we define it according to

$$\epsilon_{ijk} = \begin{cases} 1 & \text{if } i\,j\,k \text{ are an even permutation of } 1\,2\,3, \\ -1 & \text{if } i\,j\,k \text{ are an odd permutation of } 1\,2\,3, \\ 0 & \text{if any two of } i, j, k \text{ are the same.} \end{cases} \qquad (1.11)$$

We note that ϵ_{ijk} changes sign if any two of its indices are interchanged. For example, if the 1 and 3 are interchanged, then the sequence 1 2 3 becomes 3 2 1. Accordingly, we define the cross product $\boldsymbol{u} \times \boldsymbol{v}$ according to its ith component, namely,

$$(\boldsymbol{u} \times \boldsymbol{v})_i \equiv \epsilon_{ijk}u_jv_k, \qquad (1.12)$$

or, equivalently,

$$\boldsymbol{u} \times \boldsymbol{v} = (\boldsymbol{u} \times \boldsymbol{v})_i\hat{\boldsymbol{e}}_i = \epsilon_{ijk}\hat{\boldsymbol{e}}_iu_jv_k = -(\boldsymbol{v} \times \boldsymbol{u}). \qquad (1.13)$$

It is observed that this structure is closely connected to the definition of the determinant of a 3×3 matrix, which emerges from expressing the scalar triple product

$$\boldsymbol{u} \cdot (\boldsymbol{v} \times \boldsymbol{w}) = \epsilon_{ijk} u_i v_j w_k, \qquad (1.14)$$

and, by virtue of the cyclic permutivity of the Levi-Civita symbol, demonstrates that

$$\boldsymbol{u} \cdot (\boldsymbol{v} \times \boldsymbol{w}) = \boldsymbol{v} \cdot (\boldsymbol{w} \times \boldsymbol{u}) = \boldsymbol{w} \cdot (\boldsymbol{u} \times \boldsymbol{v}). \qquad (1.15)$$

The right-hand side of Eq. (1.14) is the determinant of a matrix whose rows correspond to $\boldsymbol{u}$, $\boldsymbol{v}$, and $\boldsymbol{w}$.

Indicial notation facilitates the calculation of quantities such as the vector triple cross product

$$\begin{aligned}
\boldsymbol{u} \times (\boldsymbol{v} \times \boldsymbol{w}) &= \boldsymbol{u} \times \epsilon_{ijk} \hat{\boldsymbol{e}}_i v_j w_k = \epsilon_{lmi} \hat{\boldsymbol{e}}_l u_m \epsilon_{ijk} v_j w_k \\
&= (\epsilon_{ilm} \epsilon_{ijk}) \hat{\boldsymbol{e}}_l u_m v_j w_k. \qquad (1.16)
\end{aligned}$$

It is necessary to deal first with the $\epsilon_{ilm} \epsilon_{ijk}$ term. Observe, as we sum over the i index, that contributions can emerge only if $l \neq m$ and $j \neq k$. If these conditions both hold, then we get a contribution of 1 if $l = j$ and $m = k$, and a contribution of -1 if $l = k$ and $m = j$. Hence, it follows that

$$\epsilon_{ilm} \epsilon_{ijk} = \delta_{\ell j} \delta_{mk} - \delta_{lk} \delta_{mj}. \qquad (1.17)$$

Returning to (1.16), we obtain

$$\begin{aligned}
\boldsymbol{u} \times (\boldsymbol{v} \times \boldsymbol{w}) &= (\delta_{lj} \delta_{mk} - \delta_{lk} \delta_{mj}) \hat{\boldsymbol{e}}_l u_m v_j w_k \\
&= \hat{\boldsymbol{e}}_l v_l u_m w_m - \hat{\boldsymbol{e}}_l w_l u_m v_m \\
&= \boldsymbol{v}(\boldsymbol{u} \cdot \boldsymbol{w}) - \boldsymbol{w}(\boldsymbol{u} \cdot \boldsymbol{v}), \qquad (1.18)
\end{aligned}$$

thereby reproducing a familiar, albeit otherwise cumbersome to derive, algebraic identity. Finally, if we replace the role of $\boldsymbol{u}$ in the triple scalar product (1.18) by $\boldsymbol{v} \times \boldsymbol{w}$, it immediately follows that

$$\begin{aligned}
(\boldsymbol{v} \times \boldsymbol{w}) \cdot (\boldsymbol{v} \times \boldsymbol{w}) &= |\boldsymbol{v} \times \boldsymbol{w}|^2 = \epsilon_{ijk} v_j w_k \epsilon_{ilm} v_l w_m \\
&= (\delta_{jl} \delta_{km} - \delta_{jm} \delta_{kl}) v_j w_k v_l w_m \\
&= v^2 w^2 - (\boldsymbol{v} \cdot \boldsymbol{w})^2 = v^2 w^2 \sin^2 \theta, \quad (1.19)
\end{aligned}$$

where we have made use of the definition for the angle θ given in (1.5).

The Kronecker δ and Levi-Civita ϵ permutation symbols simplify the calculation of many other vector identities, including those with respect to *derivative* operators. We define ∂_i according to

$$\partial_i \equiv \frac{\partial}{\partial x_i}, \tag{1.20}$$

and employ it to define the gradient operator ∇, which is itself a vector:

$$\nabla = \partial_i \hat{e}_i. \tag{1.21}$$

Another notational shortcut is to employ a subscript of ", i" to denote a derivative with respect to x_i; importantly, a comma "," is employed together with the subscript to designate differentiation. Hence, if f is a scalar function of $\boldsymbol{x}$, we write

$$\frac{\partial f}{\partial x_i} = \partial_i f = f_{,i}; \tag{1.22}$$

but if $\boldsymbol{g}$ is a vector function of $\boldsymbol{x}$, then we write

$$\frac{\partial g_i}{\partial x_j} = \partial_j g_i = g_{i,j}. \tag{1.23}$$

Higher derivatives may be expressed using this shorthand as well, for example,

$$\frac{\partial^2 g_i}{\partial x_j \partial x_k} = g_{i,jk}. \tag{1.24}$$

Then, the usual divergence and curl operators become

$$\nabla \cdot \boldsymbol{u} = \partial_i u_i = u_{i,i} \tag{1.25}$$

and

$$\nabla \times \boldsymbol{u} = \epsilon_{ijk} \hat{e}_i \partial_j u_k = \epsilon_{ijk} \hat{e}_i u_{k,j}. \tag{1.26}$$

Our derivations will employ Cartesian coordinates, primarily, since curvilinear coordinates, such as cylindrical and spherical coordinates, introduce a complication insofar as the unit vectors defining the associated directions change. However, once we have obtained the fundamental equations, curvilinear coordinates can be especially helpful in solving problems since they help capture the essential geometry of the Earth.

1.2 Cylindrical and Spherical Geometry

Two other coordinate systems are widely employed in geophysics, namely, cylindrical coordinates and spherical coordinates. As we indicated earlier, our starting point will always be the fundamental equations that we derived using Cartesian coordinates and then we will convert to coordinates that are more "natural" for solving the problem at hand. Let us begin in two dimensions with polar coordinates (r, θ) and review some fundamental results.

As usual, we relate our polar and Cartesian coordinates according to

$$x = r \cos \theta$$
$$y = r \sin \theta, \tag{1.27}$$

which can be inverted according to

$$r = \sqrt{x^2 + y^2}$$
$$\theta = \arctan (y/x). \tag{1.28}$$

Unit vectors in the new coordinates can be expressed

$$\hat{\boldsymbol{r}} = \cos \theta \hat{\boldsymbol{x}} + \sin \theta \hat{\boldsymbol{y}}$$
$$\hat{\boldsymbol{\theta}} = -\sin \theta \hat{\boldsymbol{x}} + \cos \theta \hat{\boldsymbol{y}}. \tag{1.29}$$

We recall how to obtain the various differential operations, such as the gradient, divergence, and curl, by using the chain rule of multivariable calculus. Suppose that f is a scalar function of x and y, and we wish to transform its Cartesian derivatives into derivatives with respect to polar coordinates. From the chain rule, it follows that

$$\frac{\partial f}{\partial x} = \frac{\partial r}{\partial x}\Big|_y \frac{\partial f}{\partial r}\Big|_\theta + \frac{\partial \theta}{\partial x}\Big|_y \frac{\partial f}{\partial \theta}\Big|_r = \cos \theta \frac{\partial f}{\partial r} - \frac{\sin \theta}{r} \frac{\partial f}{\partial \theta}, \tag{1.30}$$

where the vertical bar followed by a subscript designates the variable or variables that are held fixed. In like fashion, we can derive

$$\frac{\partial f}{\partial y} = \sin \theta \frac{\partial f}{\partial r} + \frac{\cos \theta}{r} \frac{\partial f}{\partial \theta}. \tag{1.31}$$

Finally, we can obtain the *Laplacian* of a scalar quantity in two dimensions, ∇^2, defined according to

$$\nabla^2 f \equiv \boldsymbol{\nabla} \cdot \boldsymbol{\nabla} f = \frac{\partial^2 f}{\partial x^2} + \frac{\partial^2 f}{\partial y^2} = \frac{1}{r} \frac{\partial}{\partial r}\left(r \frac{\partial f}{\partial r}\right) + \frac{1}{r^2} \frac{\partial^2 f}{\partial \theta^2}. \tag{1.32}$$

Integrals in two dimensions require the transformation of differential area elements from $dx\,dy$ to $r\,d\theta\,dr$. Therefore, the integral of f over some area A can be expressed equivalently as

$$\int_A f(x,y)\,dx\,dy = \int_A f(x_1, x_2)\,dx_1\,dx_2 = \int_A f(\boldsymbol{x})\,dA$$
$$= \int_A f(\boldsymbol{x})\,d^2x = \int_A f(r,\theta)r\,dr\,d\theta, \quad (1.33)$$

where the areas of integration are kept the same and integration over two variables is implicit.

We now move on to review three-dimensional geometry wherein polar coordinates become either cylindrical or spherical polar coordinates. We begin with cylindrical coordinates, which now introduce the third or z dimension. Accordingly, we observe that the Laplacian becomes

$$\nabla^2 f = \frac{1}{r}\frac{\partial}{\partial r}\left(r\frac{\partial f}{\partial r}\right) + \frac{1}{r^2}\frac{\partial^2 f}{\partial \theta^2} + \frac{\partial^2 f}{\partial z^2}, \quad (1.34)$$

where we assume that r is measured in the x-y plane, that is, it is not the radial distance from the origin to the point in question. Suppose, as before, that $\boldsymbol{g}$ is a vector function and we wish to obtain its divergence and curl. We will designate its components in the cylindrical coordinate system (r,θ,z) by (g_r, g_θ, g_z). These can be calculated directly by taking projections of $(f_1, f_2, f_3) = (f_x, f_y, f_z)$ onto the (r,θ,z) directions. The z direction requires no elaboration. However, we note that

$$g_r = \cos\theta g_x + \sin\theta g_y$$
$$g_\theta = -\sin\theta g_x + \cos\theta g_y \quad (1.35)$$

and

$$g_x = \cos\theta g_r - \sin\theta g_\theta$$
$$g_y = \sin\theta g_r + \cos\theta g_\theta. \quad (1.36)$$

With these results in hand, we can show that the divergence of $\boldsymbol{g}$ becomes

$$\nabla \cdot \boldsymbol{g} = \frac{1}{r}\frac{\partial}{\partial r}(r g_r) + \frac{1}{r}\frac{\partial g_\theta}{\partial \theta} + \frac{\partial g_z}{\partial z}, \quad (1.37)$$

where $\hat{\boldsymbol{r}}$, $\hat{\boldsymbol{\theta}}$, and $\hat{\boldsymbol{z}}$ are unit vectors in the associated directions. Similarly, we write the curl of $\boldsymbol{g}$ as

$$\nabla \times \boldsymbol{g} = \frac{1}{r} \left\{ \left[\frac{\partial g_z}{\partial \theta} - \frac{\partial (r g_\theta)}{\partial z} \right] \hat{\boldsymbol{r}} + \left[\frac{\partial g_r}{\partial z} - \frac{\partial g_z}{\partial r} \right] r \hat{\boldsymbol{\theta}} \right.$$
$$\left. + \left[\frac{\partial (r g_\theta)}{\partial r} - \frac{\partial g_r}{\partial \theta} \right] \hat{\boldsymbol{z}} \right\}. \quad (1.38)$$

Finally, the integral over some volume V of a scalar function f can be written equivalently as

$$\int_V f(x, y, z)\, \mathrm{d}x\, \mathrm{d}y\, \mathrm{d}z = \int_V f(x_1, x_2, x_3)\, \mathrm{d}x_1\, \mathrm{d}x_2\, \mathrm{d}x_3$$
$$= \int_V f(\boldsymbol{x})\, \mathrm{d}V = \int_V f(\boldsymbol{x})\, \mathrm{d}^3 x$$
$$= \int_V f(r, \theta, z) r\, \mathrm{d}r\, \mathrm{d}\theta\, \mathrm{d}z. \quad (1.39)$$

This concludes our summary of cylindrical coordinates.

We now adopt spherical coordinates (Figure 1.1) according to

$$x = r \sin \theta \cos \varphi$$
$$y = r \sin \theta \sin \varphi$$
$$z = r \cos \theta, \quad (1.40)$$

which can be inverted according to

$$r = \sqrt{x^2 + y^2 + z^2}$$
$$\theta = \arccos(z/r) = \arccos\left(z / \sqrt{x^2 + y^2 + z^2} \right)$$
$$\varphi = \arctan(y/x). \quad (1.41)$$

Without elaboration, we list here some essential results.

1. Unit vector relationships, from which g_r, g_θ, and g_φ can also be extracted:

$$\hat{\boldsymbol{r}} = \sin \theta \cos \varphi \hat{\boldsymbol{x}} + \sin \theta \sin \varphi \hat{\boldsymbol{y}} + \cos \theta \hat{\boldsymbol{z}}$$
$$\hat{\boldsymbol{\theta}} = \cos \theta \cos \varphi \hat{\boldsymbol{x}} + \cos \theta \sin \varphi \hat{\boldsymbol{y}} - \sin \theta \hat{\boldsymbol{z}}$$
$$\hat{\boldsymbol{\varphi}} = -\sin \varphi \hat{\boldsymbol{x}} + \cos \varphi \hat{\boldsymbol{y}}. \quad (1.42)$$

We note, as a check, that all three of these unit vectors are of unit length and are mutually orthogonal.

2. Gradient of a scalar f:

$$\nabla f = \frac{\partial f}{\partial r} \hat{\boldsymbol{r}} + \frac{1}{r} \frac{\partial f}{\partial \theta} \hat{\boldsymbol{\theta}} + \frac{1}{r \sin \theta} \frac{\partial f}{\partial \varphi} \hat{\boldsymbol{\varphi}}. \quad (1.43)$$

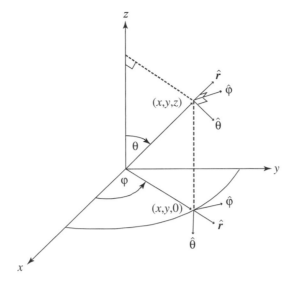

Figure 1.1.   Spherical coordinates.

3. Laplacian of a scalar f:

$$\nabla^2 f = \frac{1}{r^2 \sin \theta} \left\{ \sin \theta \frac{\partial}{\partial r} \left(r^2 \frac{\partial f}{\partial r} \right) \right.$$
$$\left. + \frac{\partial}{\partial \theta} \left(\sin \theta \frac{\partial f}{\partial \theta} \right) + \frac{1}{\sin \theta} \frac{\partial^2 f}{\partial \varphi^2} \right\}. \quad (1.44)$$

Note that the Laplacian of vector quantities will differ from the above due to the dependence of the projected components on the coordinates.

4. Divergence of a vector $\boldsymbol{g}$:

$$\nabla \cdot \boldsymbol{g} = \frac{1}{r^2 \sin \theta} \left[\sin \theta \frac{\partial (r^2 g_r)}{\partial r} + r \frac{\partial (\sin \theta g_\theta)}{\partial \theta} + r \frac{\partial g_\varphi}{\partial \varphi} \right].$$
$$(1.45)$$

5. Curl of a vector $\boldsymbol{g}$:

$$\nabla \times \boldsymbol{g} = \frac{1}{r^2 \sin \theta} \left\{ \left[\frac{\partial (r \sin \theta g_\varphi)}{\partial \theta} - \frac{\partial (r g_\theta)}{\partial \varphi} \right] \hat{\boldsymbol{r}} \right.$$
$$+ \left[\frac{\partial g_r}{\partial \varphi} - \frac{\partial (r \sin \theta g_\varphi)}{\partial r} \right] r \hat{\boldsymbol{\theta}}$$
$$\left. + \left[\frac{\partial (r g_\theta)}{\partial r} - \frac{\partial g_r}{\partial \theta} \right] r \sin \theta \hat{\boldsymbol{\varphi}} \right\}. \quad (1.46)$$

6. Volume integral of a scalar f:

$$\int_V f(\boldsymbol{x}) \, \mathrm{d}^3 x = \int_V f(r, \theta, \varphi) r^2 \sin \theta \, \mathrm{d}r \, \mathrm{d}\theta \, \mathrm{d}\varphi. \quad (1.47)$$

We will now review some of the integral relations involving vector quantities.

1.3 Theorems of Gauss, Green, and Stokes

We wish to present some familiar results from integral calculus. We will not provide proofs but will present a brief sketch as to how they can be obtained. In Figure 1.2, we depict the relevant geometry.

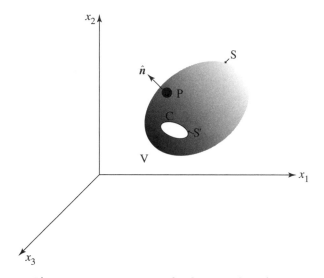

Figure 1.2.   Geometry of volume and surface.

We denote by V the volume under consideration, and S denotes the surface of that volume. We identify a point P on the surface of that volume, and show by an arrow the unit vector $\hat{\boldsymbol{n}}$ emerging out from that surface. Finally, we draw a closed curve C on that surface that contains a surface area S'. We denote by $\boldsymbol{g}$ a vector function and by f and h two different scalar functions. We assume that f, $\boldsymbol{g}$, and h all go to zero as our distance from the origin goes to infinity. As before, we denote surface and volume elements by d^2x and d^3x, respectively.

Gauss's theorem can be expressed by

$$\int_V \nabla \cdot \boldsymbol{g}\, d^3x = \int_S \boldsymbol{g} \cdot \hat{\boldsymbol{n}}\, d^2x. \tag{1.48}$$

This result can be proved by subdividing the volume V into a set of cubes, going to the limit that the sides of the cubes become vanishingly small. It is simple to show by a direct integration that Gauss's theorem holds for each cube. When we amalgamate all of the cubes, the contributions emerging from the surfaces that are in common cancel, leaving only the contribution from the external enveloping surface.

Green's theorem (Morse and Feshbach, 1999) is generally expressed as

$$\int_V \nabla \cdot (f\nabla h - h\nabla f)\,\mathrm{d}^3 x = \int_S (f\nabla h - h\nabla f) \cdot \hat{\boldsymbol{n}}\,\mathrm{d}^2 x$$

$$= \int_V (f\nabla^2 h - h\nabla^2 f)\,\mathrm{d}^3 x. \quad (1.49)$$

Other textbooks, for example, Greenberg (1998), refer to this as one of Green's identities. The second integral is a direct application of Gauss's theorem. To obtain the third integral, we applied the $\nabla\cdot$ operator on the product of the two terms and eliminated the common $\nabla f \cdot \nabla h$ term, expressing $\nabla \cdot \nabla$ as the Laplacian ∇^2.

Stokes's theorem allows us to relate the integral of the curl of a vector acting upon the surface S' that is enclosed by a curve C to the integral of that vector projected onto and along that curve C. It emerges in electromagnetic theory and has the form

$$\int_{S'} (\nabla \times \boldsymbol{g}) \cdot \hat{\boldsymbol{n}}\,\mathrm{d}^2 x = \oint_C \boldsymbol{g} \cdot \mathrm{d}\boldsymbol{\ell}, \quad (1.50)$$

where $\oint$ denotes an integral around a closed curve, in this case C, and $\mathrm{d}\boldsymbol{\ell}$ is a differential line element that resides on C. As in the case of Gauss's theorem, we can prove Stokes's theorem by subdividing the area enclosed by the curve into a set of squares whose sides will ultimately be taken to be vanishingly small. Stokes's theorem can readily be proven on a square, and the lines in common among the squares cancel when calculating the contribution from all squares in the limit.

1.4 Rotation and Matrix Representation

We have already explored one form of vector rotation, namely, the conversion of Cartesian coordinates into spherical geometry where the unit vectors $\hat{\boldsymbol{x}}$, $\hat{\boldsymbol{y}}$, and $\hat{\boldsymbol{z}}$ were replaced by or "rotated" into $\hat{\boldsymbol{r}}$, $\hat{\boldsymbol{\theta}}$, and $\hat{\boldsymbol{\varphi}}$. We wish to obtain a general expression for converting coordinates from a system of axes $\hat{\boldsymbol{e}}_1$, $\hat{\boldsymbol{e}}_2$, and $\hat{\boldsymbol{e}}_3$ to new coordinate axes $\hat{\boldsymbol{e}}_1'$, $\hat{\boldsymbol{e}}_2'$, and $\hat{\boldsymbol{e}}_3'$. (We assume that you have had an introductory linear algebra course as an undergraduate, including the solution of linear equations via Gaussian elimination and introduction to the eigenvalue problem.) The visual approach to coordinate conversions as well as rotation is typically presented diagrammatically in two dimensions, but becomes rather cumbersome and confusing in three. Matrix algebra provides a simple way of clarifying this problem. We recognize that the new coordinate axes $\hat{e}'_i$ for $i = 1, 2, 3$ should be expressible as a linear combination of the $\hat{e}_j$ coordinate axes. Suppose that A is a 3×3 matrix with components A_{ij} so that we can write

$$\hat{e}'_i = \sum_{j=1}^{3} A_{ij}\hat{e}_j = A_{ij}\hat{e}_j \tag{1.51}$$

for $i = 1, 2$, and 3, returning to indicial notation, since this is a general expression for a linear combination of the original coordinate axes. Accordingly, we take the dot product of this expression with $\hat{e}_k$, for $i, k = 1, \ldots, 3$ and observe that

$$\hat{e}_k \cdot \hat{e}'_i = \hat{e}'_i \cdot \hat{e}_k = \sum_{j=1}^{3} A_{ij}\hat{e}_k \cdot \hat{e}_j = A_{ik} \tag{1.52}$$

by virtue of the Kronecker δ identity (1.7). A useful mnemonic device for remembering this result is that A_{ik} is the matrix that connects the kth axis from the original coordinate to the ith axis in the new coordinate system. Similarly, we introduce another matrix B so that we can write the inverse operation, going from the new coordinates back to the old, namely,

$$\hat{e}_i = \sum_{i=1}^{3} B_{ij}\hat{e}'_j = B_{ij}\hat{e}'_j, \tag{1.53}$$

returning to indicial notation, and directly obtain that

$$B_{ik} = \hat{e}_i \cdot \hat{e}'_k. \tag{1.54}$$

We observe that

$$B_{ik} = A_{ki}, \tag{1.55}$$

establishing that the matrix $B = \tilde{A}$ where the tilde ~ designates the *transpose* of the matrix. Hence, it immediately follows that

$$\tilde{A}A = A\tilde{A} = I, \tag{1.56}$$

where I designates the identity matrix so that its ij components are δ_{ij}. For self-evident reasons, we refer to both A and B as *rotation matrices*.

The conversion of the coordinates of a point from one coordinate system to another, which is a rotated version (without translation) of the original, can also be regarded as a rotation in the opposing sense of the point undergoing coordinate conversion. For example, suppose our coordinate conversion corresponded to a positive 45° rotation of the x-y axes to a new x'-y' set of axes, leaving the z-axis unchanged. That would have the same effect as a rotation of negative 45° in the x-y plane of the point under consideration.

Returning momentarily to the expression $A_{ij} = \hat{e}'_i \cdot \hat{e}_j$, it follows that each of the matrix components is a *direction cosine*, the inner product of the original jth unit vector with the new ith unit vector. Had we attempted to do this using a visual construction, for example, as performed by Goldstein et al. (2002), we would have observed the same relationships, albeit in a geometrically complicated form. One final feature of rotation matrices that is often important is that the three rows of a rotation matrix, if we were to regard them as three vectors, are mutually orthogonal and of unit length. A similar observation can be made for the three columns of a rotation matrix.

Suppose now that we have a vector v that we wish to express in both coordinate systems utilizing (1.1) and indicial notation, namely,

$$v = v_i \hat{e}_i \equiv v' = v'_j \hat{e}'_j. \tag{1.57}$$

Taking the inner product of $\hat{e}'_k$ with this expression, we observe that

$$v'_k = v_j \hat{e}'_k \cdot \hat{e}_j = \hat{e}'_k \cdot \hat{e}_j v_j = A_{kj} v_j. \tag{1.58}$$

Therefore, we note that the Cartesian coordinate representation in the new reference (rotated) frame transforms in the same way as the coordinate axes transform. For more insight into the nature of rotations, consult Goldstein et al. (2002) or Newman (2012).

Before proceeding, it is useful to describe two further aspects of matrix structure. Any matrix A can be expressed as the sum of a symmetric and an antisymmetric matrix, namely,

$$A = \frac{A + \tilde{A}}{2} + \frac{A - \tilde{A}}{2}, \tag{1.59}$$

where $\tilde{A}$ designates the transpose of the matrix. Rotation matrices have a special structure, inasmuch as they possess both a symmetric and an antisymmetric part. Intuitively, we expect that a rotation matrix identifies a special direction corresponding to the axis around which the rotation takes place, and the angle of rotation that is executed around this axis. It takes two variables to identify the direction—think of them as corresponding to a latitude and longitude—and a third variable to identify the magnitude of the rotation. This situation applies whether we are considering a physical rotation of an object or the expression of its orientation in a new coordinate system. Newman (2012) provides a more detailed discussion of this problem. There are other methodologies for describing a rotation, such as through three *Euler angles*. Goldstein et al. (2002) provides a detailed discussion of this approach. The Euler angles are routinely employed in celestial mechanics in describing the orientation of elliptical orbits in the gravitational two-body problem, for example, for cometary orbits (Roy, 2005).

We consider an x-y plane established by the orbit of Jupiter around the Sun, with the x-direction corresponding to the direction of Jupiter's perihelion; we wish to establish the orientation of the comet's elliptical orbit with respect to this *invariable plane*. To do this, we undertake a series of rotations of the comet's orbit. We begin with the ellipse being in the x-y plane with the Sun at the origin, and the ellipse's axis initially oriented in the x-direction. We then rotate the ellipse in the x-y plane, that is, around the z-axis, by an angle referred to as the *longitude of the ascending node Ω*. The *line of nodes* identifies the y-axis about which the plane of the orbit is now rotated in the x-z plane to corresponds to its *angle of inclination i*. Finally, we perform a third rotation, this time around the new z-axis in the x-y plane to identify the *longitude of perihelion ω*. For details of this procedure, the reader is encouraged to consult Goldstein et al. (2002), Roy (2005), Taff (1985), and Danby (1988). Importantly, what should be evident is that, by executing three successive rotations around the z, y, and z axes, respectively, it is possible to orient properly any three-dimensional object. There are alternative conventions for the Euler angles that are detailed in Goldstein et al. (2002). We have focused here upon the one commonly employed in planetary science as well as in classical mechanics. Rotation matrices occupy an important role in describing the dynamics of objects and materials in many different environments.

A natural question to ask now is, how do matrices derived from physically based considerations such as a variational or energy principle transform under coordinate axis rotation? This is the defining characteristic that distinguishes tensors from matrices.

1.5   Tensors, Eigenvalues, and Eigenvectors

In this section, we will explore how we can make symmetric 3×3 matrices transform under coordinate rotation in the same way as a vector; in so doing, we refer to the matrix as being a second-rank tensor. In addition, we observe that there exists a special coordinate basis where the action of the tensor upon a vector is equivalent to a scalar multiplication. This simplifies many calculations, and eigenvalue-eigenvector analysis is at the heart of this procedure. We will demonstrate how to solve the eigenvalue problem, and show how the eigenvectors form the basis set for the new coordinate system.

Having observed (1.57) describing how vectors transform, namely,

$$v_i' = A_{ij}v_j = \tilde{B}_{ji}v_j, \tag{1.60}$$

we introduce the concept of a second-rank tensor $\boldsymbol{T}$ as being a matrix $\boldsymbol{T}$ with components $T_{k\ell}$ that transforms similarly, namely,

$$T_{ij}' = A_{ik}A_{j\ell}T_{k\ell} = A_{ik}T_{k\ell}\tilde{A}_{\ell j} = \tilde{B}_{ik}T_{k\ell}B_{\ell j} \tag{1.61}$$

and

$$T_{ij} = B_{ik}B_{j\ell}T_{k\ell}' = B_{ik}T_{k\ell}'\tilde{B}_{\ell j} = \tilde{A}_{ik}T_{k\ell}A_{\ell j}. \tag{1.62}$$

Accordingly, we write

$$\boldsymbol{T}' = \boldsymbol{A}\boldsymbol{T}\tilde{\boldsymbol{A}} \quad \text{or} \quad \boldsymbol{T} = \tilde{\boldsymbol{A}}\boldsymbol{T}'\boldsymbol{A}, \tag{1.63}$$

which is routinely referred to as a *similarity transformation.* (Equivalent expressions are available utilizing $\boldsymbol{B}$.)

Matrix quantities in geophysics are generally real valued and frequently *symmetric*; if A is symmetric, we say that

$$A_{ij} = A_{ji}. \tag{1.64}$$

Tensors are often associated with situations where there are special associated directions. For example, in geophysics and engineering applications, the stress and strain tensors describe how compressional or extensional forces act upon materials resulting in a displacement according to a generalization of Hooke's law. In particular, if T is a tensor and u is a vector, we can find a number λ such that $T \cdot u = \lambda u$. In this situation, we refer to λ as an *eigenvalue* or *characteristic value* of T and u is its associated *eigenvector* or *characteristic vector*. For the examples mentioned, the stress and strain tensors are symmetric, thereby guaranteeing the existence of real eigenvalues and eigenvectors. Moreover, there are special directions where the force associated with a solid or plastic material emerges in a direction orthogonal or normal to a surface (Newman, 2012). Similarly, in rotational kinematics (see, e.g., Goldstein et al., 2002) the moment of inertia tensor has special directions associated with it, often as an outcome of symmetry considerations. Since most geophysics graduate students have already completed an analytical mechanics course, but possibly not continuum mechanics, we shall explore rotational motion as an example of the eigenvalue problem.

Suppose we wish to calculate the kinetic energy K of a rigid solid body rotating around its center of mass with angular velocity $\boldsymbol{\omega}$, where the direction of this vector corresponds to the orientation of the spin axis. It follows, for any point x inside the rotating body, that the corresponding velocity is

$$v = \boldsymbol{\omega} \times x \tag{1.65}$$

and the kinetic energy satisfies

$$K = \frac{1}{2} \int \rho(x) v^2(x) \, \mathrm{d}^3 x. \tag{1.66}$$

Importantly, we note that this is an integral over non-negative quantities and must necessarily yield a positive result. After some brief algebra, we find that

$$K = \frac{1}{2} \int \rho(x)[x_k x_k \omega_i \delta_{ij} \omega_j - \omega_i \omega_j x_i x_j] \, \mathrm{d}^3 x$$

$$= \omega_i I_{ij} \omega_j, \tag{1.67}$$

which we refer to as a *quadratic form* where the components of the moment of inertia tensor I_{ij} satisfy

$$I_{ij} = \int \rho(\boldsymbol{x})[x_k x_k \delta_{ij} - x_i x_j]. \tag{1.68}$$

We immediately note that this tensor is real and symmetric.

Tensor analysis, especially the properties of eigenvalues and eigenvectors, is an important topic generally treated in advanced undergraduate mathematics courses. However, there are some features of eigenvalue analysis for tensorial quantities that are conceptually vital in geophysics, and we sketch here some of their properties. More complete physically motivated treatments can be found in Goldstein et al. (2002) and in Newman (2012), which show that 3×3 symmetric tensors have real-valued eigenvalues and eigenvectors. In this particular case, the eigenvalues must be positive (or possibly zero) since the kinetic energy can never be negative. The eigenvalues, in turn, are the solutions to the cubic (characteristic) polynomial $p(\lambda)$ constructed by solving for the three roots of the equation

$$p(\lambda) = \det(\boldsymbol{A} - \lambda \boldsymbol{I}) = 0. \tag{1.69}$$

(Please note that we are employing I_{ij} to designate the components of the moment of inertia tensor. Many linear algebra textbooks employ the same symbol to describe the components of the identity matrix.) For this physically important class of eigenvalue problems, Cardano's method (Newman, 2012) provides an explicit closed-form solution, and we now sketch the derivation of this method.

Suppose the cubic polynomial can be expressed

$$a\lambda^3 + b\lambda^2 + c\lambda + d = 0, \tag{1.70}$$

where $a \neq 0$. We choose to eliminate the quadratic term by replacing λ with $z + \alpha$, where $\alpha = -b/3a$. We now obtain the equation

$$az^3 + c'z + d' = 0, \tag{1.71}$$

where $c' = c + 2\alpha b + 3\alpha^2 a$ and $d' = d + c\alpha + b\alpha^2 + a\alpha^3$. We now exploit the *triple-angle identity*

$$\cos^3 \theta = \tfrac{1}{4} \cos 3\theta + \tfrac{3}{4} \cos \theta, \tag{1.72}$$

which can be readily verified by writing $\cos\theta$ in exponential form. We replace z in (1.71) by $y\cos\theta$ and then obtain

$$y\cos\theta\left[\frac{3ay^2}{4} + c'\right] + \left[\frac{ay^3\cos 3\theta}{4} + d'\right] = 0. \qquad (1.73)$$

We are at liberty to select y so that the first bracketed term disappears, and we then select θ so that the second bracketed term disappears. Note, however, that the solution for θ is *degenerate*; upon finding one solution, we can construct two additional solutions that are also real valued by taking our original solution and adding $2\pi/3$ as well as taking our original solution and subtracting $2\pi/3$. This kind of degeneracy is common in complex analysis and is associated with *branch cuts*, a topic we will review in chapter 3. Remarkably, while often overlooked in most textbooks, real-world, three-dimensional eigenvalue solutions are readily within reach.

Finally, we note that for each of the eigenvalues, we can now explicitly calculate the eigenvectors by solving the associated pair of linear equations and then normalizing the vectors obtained to be of unit length. We designate, after ordering, the eigenvalues represented by I_i according to

$$0 \leqslant I_1 \leqslant I_2 \leqslant I_3. \qquad (1.74)$$

(In planetary physics, the eigenvalues are often referred to as $0 \leqslant A \leqslant B \leqslant C$.) We now define the normalized eigenvectors associated with the three eigenvalues as $\hat{e}'_\ell$, for $\ell = 1, 2$, and 3. If this selection does not yield a right-handed coordinate basis, we typically change the sign of one of the eigenvectors to conform with that convention. In terms of our original coordinate basis, the coordinates of the ℓth eigenvector $\hat{u}_\ell$, which we now equate with $\hat{e}'_\ell$, can be expressed as column vectors

$$\hat{u}_\ell = \begin{pmatrix} \hat{e}_1 \cdot \hat{e}'_\ell \\ \hat{e}_2 \cdot \hat{e}'_\ell \\ \hat{e}_3 \cdot \hat{e}'_\ell \end{pmatrix}. \qquad (1.75)$$

Taken together, we recognize that the three column vectors (for $\ell = 1, 2, 3$) establish the rotation matrix $\tilde{A}$, namely,

$$\tilde{A} = \begin{pmatrix} \hat{e}_1 \cdot \hat{e}'_1 & \hat{e}_1 \cdot \hat{e}'_2 & \hat{e}_1 \cdot \hat{e}'_3 \\ \hat{e}_2 \cdot \hat{e}'_1 & \hat{e}_2 \cdot \hat{e}'_2 & \hat{e}_2 \cdot \hat{e}'_3 \\ \hat{e}_3 \cdot \hat{e}'_1 & \hat{e}_3 \cdot \hat{e}'_2 & \hat{e}_3 \cdot \hat{e}'_3 \end{pmatrix}. \qquad (1.76)$$

It is now easy to verify that

$$T\tilde{A} = \tilde{A}\Lambda, \tag{1.77}$$

where Λ is the diagonal matrix containing the eigenvalues λ_i or, equivalently, I_i, namely,

$$\Lambda = \begin{pmatrix} \lambda_1 & 0 & 0 \\ 0 & \lambda_2 & 0 \\ 0 & 0 & \lambda_3 \end{pmatrix}. \tag{1.78}$$

Multiplying each side of the equation from the right by A, we then recover the similarity transform (1.61). In the important special case of stress or strain tensors, which are symmetric, a similar analysis is applicable where their eigenvalues are necessarily real valued, but can have either sign. A similar analysis is applicable to stress and strain tensors (although their eigenvalues need not be non-negative).

1.6 *Ramp, Heaviside, and Dirac δ Functions*

We now wish to introduce the concept of a *generalized function*, a function on the real line that vanishes essentially everywhere but the origin, where it is singular, and has an integral of 1. This concept was introduced by Paul Dirac and is the continuous analogue of the Kronecker δ_{ij} that we introduced earlier. In particular, we define $\delta(x)$ according to

$$\delta(x) = 0 \quad \text{if } x \neq 0$$

$$g(y) = \int_{-\infty}^{\infty} g(x)\delta(x - y)\,dx \tag{1.79}$$

for any $g(x)$. Consider the integral over the δ function and we define the step or *Heaviside* function according to

$$H(x) = \int_{-\infty}^{x} \delta(y)\,dy$$

$$= \begin{cases} 0 & \text{if } x < 0, \\ 1 & \text{if } x > 0. \end{cases} \tag{1.80}$$

Additionally, we introduce the *ramp function* $R(x) \equiv xH(x)$ and observe that it satisfies

$$R(x) = \int_{-\infty}^{x} H(y)\,dy. \tag{1.81}$$

With these expressions, we also observe that

$$\delta(x) = \frac{dH(x)}{dx}$$

$$H(x) = \frac{dR(x)}{dx}$$

$$\delta(x) = \frac{d^2 R(x)}{dx^2}. \qquad (1.82)$$

We plot these functions in Figure 1.3.

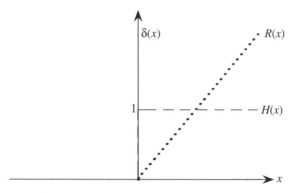

Figure 1.3. Dirac $\delta(x)$, Heaviside $H(x)$, and ramp $R(x)$ functions. Here, the δ function (solid line) vanishes everywhere but the origin, where it is singular. The Heaviside function (dashed line) vanishes for negative values, while the ramp function (dotted line) also vanishes for negative values.

The fundamental aspect present here is that the derivative of a discontinuous function yields a δ function while the second derivative of a function with discontinuous derivative does the same. The quantitative jump in function values or derivatives controls the amplitude of the outcome. We will have much more to say about this topic and its relation to *Green's functions* in later chapters.

Having provided this simple digest of mathematical preliminaries, with an emphasis upon geometry, we turn in the next chapter to a brief discussion of some of the problems expressible via partial differential equations that are central to contemporary geophysics.

1.7 Exercises

1. Prove the following identities using the Kronecker δ and Levi-Civita ϵ permutation symbol identities.

(a) Show that
$$(\boldsymbol{a} \times \boldsymbol{b}) \cdot \boldsymbol{a} = 0.$$

(b) Show that
$$\nabla \times (\nabla \times \boldsymbol{u}) = \nabla(\nabla \cdot \boldsymbol{u}) - \nabla^2 \boldsymbol{u}.$$

2. Suppose that $\boldsymbol{b}$ is an arbitrary point in 3-space. Let $\mathbb{X}$ be the set of points $\boldsymbol{x}$ such that $(\boldsymbol{x} - \boldsymbol{b}) \cdot \boldsymbol{x} = 0$. Show that this describes the surface of a sphere with center $\frac{1}{2}\boldsymbol{b}$ and radius $\frac{1}{2}b$.

3. Derive using the chain rule the formula for the Laplacian in cylindrical coordinates.

4. Derive using the chain rule the formula for the Laplacian in spherical coordinates.

5. Let $\boldsymbol{w}(\boldsymbol{x}, t)$, called the vorticity of a flow, be defined by
$$\boldsymbol{w}(\boldsymbol{x}, t) \equiv \nabla \times \boldsymbol{v}(\boldsymbol{x}, t).$$

Suppose that our flow corresponds to solid-body rotation, that is,
$$\boldsymbol{v}(\boldsymbol{x}, t) = \boldsymbol{\Omega} \times \boldsymbol{x},$$

where $\boldsymbol{\Omega}$ is a constant vector that describes the rotation rate of the object. Prove that
$$\boldsymbol{w} = 2\boldsymbol{\Omega}.$$

6. Show that the divergence of the velocity field in the previous problem is zero. Therefore, using Gauss's theorem, prove that the flux of material traveling with this velocity through *any* surface is zero, that is, the total amount of fluid in any container satisfying these relationships is conserved. Explain *why* this is the case.

7. Consider a new (primed) set of coordinates defined by the vectors
$$\hat{\boldsymbol{e}}_1' = \begin{pmatrix} \frac{1}{\sqrt{2}} \\ \frac{1}{\sqrt{2}} \\ 0 \end{pmatrix}; \quad \hat{\boldsymbol{e}}_2' = \begin{pmatrix} -\frac{1}{\sqrt{2}} \\ \frac{1}{\sqrt{2}} \\ 0 \end{pmatrix}; \quad \hat{\boldsymbol{e}}_3' = \begin{pmatrix} 0 \\ 0 \\ 1 \end{pmatrix}.$$

(Coordinate axes of this sort can emerge in the calculation of phase diagrams in geochemistry.) Prove that they are mutually orthogonal and of unit length (orthonormal). Find the rotation matrix that transforms from the original into the new coordinates. Find the inverse transformation for this rotation.

8. Using Cardano's method as described in the text, solve for the roots of the cubic polynomial

$$f(z) = z^3 - 7z^2 + 14z - 8.$$

9. Suppose we have a matrix of the form

$$\begin{pmatrix} \frac{5}{2} & -\frac{1}{2} & 0 \\ -\frac{1}{2} & \frac{5}{2} & 0 \\ 0 & 0 & 1 \end{pmatrix}.$$

Solve for the eigenvalues of this tensor. Can this be a moment of inertia tensor? Explain why or why not. What are its eigenvectors? Express in similarity form this tensor in terms of its associated diagonal matrix and rotation matrices.

CHAPTER TWO

Ordinary Differential Equations

We will begin this chapter by reviewing some features of ordinary differential equations and their methods of solution. Ordinary differential equations emerge directly in many situations, for example, radioactive decay calculations in geochronology in the dynamics of planetary bodies, and simple models of the Earth's dynamo. What distinguishes ordinary from partial differential equations is that the former depends only upon a single independent variable, such as time in the examples mentioned above. Partial differential equations, in contrast, have two or more independent variables. Moreover, as we develop methods for solving partial differential equations in chapter 4, especially those involving Laplacians in Cartesian, cylindrical, and spherical environments, we discover that linear partial differential equations may be simplified using the method of separation of variables, which produces a suite of ordinary differential equations. In particular, Coddington and Levinson (1984) present a classic treatment of the classification and nature of ordinary differential equations with a particular focus on the nature of solutions in the vicinity of so-called "fixed points." Bender and Orszag (1999) is a modern, comprehensive treatment that delves into many approximate methods for solving ordinary differential equations, as well as chaotic behavior. Davis (1960) is less of a textbook and more of a pleasurable read on the topic of nonlinear ordinary differential as well as integral equations. Note that Jeffreys and Jeffreys (1999) provide an approach to mathematical physics that puts particular emphasis on the special functions that characterize solutions to partial differential equations involving the Laplacian operator. Whittaker and Watson (1979) also pursue that strategy, while Watson (1995) is the seminal volume addressing the nature of the Bessel function, which is central to two-dimensional as well as cylindrical geometry. Butkov (1968), albeit out of print, provides a good survey of the special functions. Strogatz (1994) provides a comprehensive survey, appropriate for introductory graduate courses, on issues attendant to nonlinear dynamics and to chaos. Nayfeh and Balachandran (1995), as well as in other volumes that Nayfeh has authored, provide an in depth almost recipe-like presentation addressing nonlinear dynamical problems, perturbation theory, and how to solve such problems. Finally, Lichtenberg and Lieberman (1992) approach the issues of nonlinear dynamics at a research level, and are well worth consulting after studying the previous two volumes.

Ordinary differential equations are fundamental to all arenas of physics, beginning with Newton's paradigmatic second law $F = ma$. We will review some elementary methods of solving ordinary differential equations, and move on to dynamical issues attendant to Newton's laws. Then, we will explore more complex systems, such as LRC circuits including, especially, the geophysical analogue of a Maxwell visco-elastic solid. We will introduce concepts associated with driven oscillators, resonance, and the method of variation of constants. Then, we will explore approximation methods—some of which originated in geophysics—such as the JWKB method for second-order ordinary differential equations and the concept of adiabatic invariants, which occupy a central role in space plasma physics. Brief mention will be given to Lagrangian and Hamiltonian dynamics and the topology of such systems, which govern problems ranging from planetary orbits to the mean behavior of ocean currents. In classical mathematical methods of physics textbooks, the chapter(s) addressing ordinary differential equations present an opportunity to introduce the method of separation of variables, which is exploited to render two- and three-dimensional geometries as a combination of one-dimensional problems that have the form of second-order ordinary differential equations. We shall do the same here and employ this opportunity to introduce Bessel functions and azimuthal modes in two dimensions and Legendre polynomials in three dimensions with the attendant development of spherical harmonics. We will use this opportunity to solve one of the fundamental problems in geophysics and planetary physics: the relationship between the mass density distribution of a planet and its potential energy function, as well as the potential's variation outside the body. This too will allow us to expand upon the nature of the Dirac δ function and Green's functions anticipated in the previous chapter. While the preceding discussion has been largely classical, this provides an entry point for the discussion of nonlinearity. Following our very brief discussion of Newtonian dynamics, we will now explore the richer topic arising from coupled ordinary differential equations, ranging from systems such as those employed to help explain the Earth's dynamo (Bullard model) and turbulence (Lorenz model). The trajectories of solutions, especially for the latter, are exceedingly rich and provide a backdrop for introducing the concepts of fractals, self-similarity, and mappings. Let us now begin this journey.

2.1 Linear First-Order Ordinary Differential Equations

Let us suppose that we wish to solve for some variable y as a function of x, that is, $y(x)$ where x and y are scalar variables. We refer to y as the *dependent variable* and x as the *independent variable*. We use the usual calculus-based framework for calculating derivatives and employ a prime "$'$" to denote a spatial derivative where x is a spatial variable. (If our independent variable is time, we employ a dot to denote the time derivative.) Hence, we have the usual notation

$$y'(x) \equiv \frac{dy(x)}{dx} \tag{2.1}$$

and designate higher derivatives accordingly,

$$y''(x) \equiv \frac{d^2 y(x)}{dx^2}$$
$$y^{(n)}(x) \equiv \frac{d^n y(x)}{dx^n}, \tag{2.2}$$

where $n \geqslant 2$ is an integer. The *order* of a differential equation, whether ordinary or partial, corresponds to the highest derivative that is present, while the *degree* corresponds to the power to which the highest-order derivative is raised, after the equation has been rationalized to contain only integer powers of derivatives. The general solution to an ordinary differential equation is the most general function $y(x)$ that satisfies the equation, and contains *constants of integration* that may be determined from some *initial condition* or, in some cases, *boundary conditions*. We shall pursue our discussion by examining a progression of ordinary differential equations.

We begin by considering the simplest linear first-order, first-degree ordinary differential equation, namely,

$$\frac{dy(x)}{dx} = b(x), \tag{2.3}$$

whose solution can also be expressed as an integral—chapter 3 will address, among other topics, approximation methods for evaluating integrals. Since this *inhomogeneous equation* is linear, a *general or complete solution* to this problem is obtained by finding a *particular solution* and adding to it the solution to the *homogeneous equation*, assuming that the right-hand side is identically zero. (In the case of higher-order linear differential equations, say of order n, there exist n different homogeneous solutions and a linear combination of them must be added to a particular solution to obtain the general solution.) The constant(s) associated with the homogeneous solutions are established by satisfying the initial condition, which, in this case, is given by

$$y(x_0) = y_0. \tag{2.4}$$

This differential equation has the immediate solution

$$y(x) = \int_{x_0}^{x} b(x')\,dx' + y_0 \tag{2.5}$$

and contains an additive constant, the so-called *constant of integration*, which we have explicitly linked to the *initial condition*. The result is given in *closed form* through the use of an integral or so-called *quadrature*. This ordinary differential equation, however, is almost trivial in that there is no term in $y(x)$ present.

Accordingly, let us consider the first-order, first-degree ordinary differential equation

$$\frac{dy(x)}{dx} + ay(x) = 0, \tag{2.6}$$

using the same initial condition as before. We have expressed our equation, including a zero on the right-hand side, to emphasize that, apart from the dependencies upon $y(x)$ and its derivative on the left-hand side, there is no "source" driving the evolution of $y(x)$. We refer to this kind of equation as being *homogeneous*.

We immediately recognize that the solution is

$$y(x) = y_0 \exp[-a(x - x_0)]. \tag{2.7}$$

This expression indicates the importance of exponentials in our calculations. We can develop a further understanding of this so-called constant coefficient problem by expressing (2.6) using operator form, namely,

$$\left\{\frac{d}{dx}+a\right\}y(x) = \exp[-a(x-x_0)]\frac{d}{dx}\{\exp[a(x-x_0)]y(x)\} = 0,$$
(2.8)

where the term $\exp[a(x - x_0)]$ is referred to as an *integrating factor*. The use of integrating factors provides an effective means of collapsing complex linear operators of arbitrarily high order (but of degree one) to a much simpler expression. We immediately observe that the leading exponential in the expression on the right can be ignored, leaving us with the solution

$$\exp[a(x - x_0)]y(x) = c,$$
(2.9)

where c is a constant that we can identify with y_0 when x is set to x_0.

We can generalize from this case to the situation where a is now a function of x, that is, $a(x)$. We therefore write

$$\frac{dy(x)}{dx} + a(x)y(x) = 0$$
(2.10)

and observe that the integration factor has now become

$$\exp\left[\int_{x_0}^{x} a(x')\,dx'\right]$$

and our operator form for (2.10) becomes

$$\exp\left[\int_{x_0}^{x} -a(x')\,dx'\right]\frac{d}{dx}\left\{\exp\left[\int_{x_0}^{x} a(x')\,dx'\right]y(x)\right\} = 0.$$
(2.11)

Once again, we observe that this equation is immediately integrable, in parallel with (2.9), and we obtain

$$y(x) = y_0 \exp\left[\int_{x_0}^{x} -a(x')\,dx'\right]$$
(2.12)

for this more general linear but homogeneous equation.

Let us now consider the inhomogeneous case, where we have a source term $b(x)$, as in (2.3), namely,

$$\frac{dy(x)}{dx} + a(x)y(x) = b(x).$$
(2.13)

Using (2.11), we can write this as

$$\exp\left[\int_{x_0}^x -a(x')\,\mathrm{d}x'\right]\frac{\mathrm{d}}{\mathrm{d}x}\left\{\exp\left[\int_{x_0}^x a(x')\,\mathrm{d}x'\right]y(x)\right\} = b(x),$$

$$\text{(2.14)}$$

which is also immediately integrable. We multiply both sides of the equation by our new integrating factor and obtain

$$\frac{\mathrm{d}}{\mathrm{d}x}\left\{\exp\left[\int_{x_0}^x a(x')\,\mathrm{d}x'\right]y(x)\right\} = b(x)\exp\left[\int_{x_0}^x a(x'')\,\mathrm{d}x''\right]$$

$$\text{(2.15)}$$

and integrate to get

$$\exp\left[\int_{x_0}^x a(x')\,\mathrm{d}x'\right]y(x)$$

$$= \int_{x_0}^x b(x')\exp\left[\int_{x_0}^{x'} a(x'')\,\mathrm{d}x''\right]\mathrm{d}x' + c', \quad \text{(2.16)}$$

where c' is a constant that is to be determined. Going to the limiting case of $x = x_0$, we observe that our constant of integration c' is, once again, our initial condition y_0. Finally, all of this can be amalgamated to give

$$y(x) = \exp\left[\int_{x_0}^x -a(x')\,\mathrm{d}x'\right]\int_{x_0}^x b(x')\exp\left[\int_{x_0}^{x'} a(x'')\,\mathrm{d}x''\right]\mathrm{d}x'$$

$$+ \exp\left[\int_{x_0}^x -a(x')\,\mathrm{d}x'\right]y_0. \quad \text{(2.17)}$$

Finally, we can combine the two exponentials in the first term to give

$$y(x) = \int_{x_0}^x b(x')\exp\left[\int_{x_0}^x -a(x'')\,\mathrm{d}x''\right]$$

$$\times \exp\left[\int_{x_0}^{x'} a(x'')\,\mathrm{d}x''\right]\mathrm{d}x'$$

$$+ \exp\left[\int_{x_0}^x -a(x')\,\mathrm{d}x'\right]y_0, \quad \text{(2.18)}$$

which reduces to

$$y(x) = \int_{x_0}^x b(x')\exp\left[-\int_{x'}^x a(x'')\,\mathrm{d}x''\right]\mathrm{d}x'$$

$$+ \exp\left[\int_{x_0}^x -a(x')\,\mathrm{d}x'\right]y_0. \quad \text{(2.19)}$$

We have now derived the general solution to the general linear first-order, first-degree ordinary differential equation.

This class of ordinary differential equation has had a prominent role in establishing the geochronology of the Earth. Suppose, now, that y represents the abundance of some isotope that decays at a rate a, where we now take x to represent the time. Let us further assume that y is the daughter nucleus of yet another decay process, and that we are witnessing a chain of decays. An excellent example of this involves the α-decay of ^{230}Th into ^{226}Ra and then into ^{222}Rn; these decay processes have half-lives of 75,380 yr and 1602 yr, respectively. This thorium $\rightarrow$ radium $\rightarrow$ radon chain involves very different decay rates. In our example, y would correspond to the radium isotope and a to its (constant) decay rate into radon (a would be approximately 0.69315 divided by its half-life). Meanwhile, b would be time dependent and would correspond to the decay rate of that isotope of thorium into radium times the prevailing abundance of thorium. Thus, we observe that equations of the type that we have just discussed occupy a prominent role in isotope geochemistry. We will revisit this issue in chapter 4 and Figure 4.4.

Linear combinations of first-order, first-degree ordinary differential equations with constant coefficients abound in the earth sciences, notably in geochemistry. Their treatment generally requires the introduction of matrices and the determination of their eigenvalues, which are intimately related to the decay rates present. We will not describe further here the mechanics of addressing such problems.

Before proceeding to second-order ordinary differential equations, we wish to introduce a topic that will ultimately become essential in dealing with *chaos*. Suppose that $x(t)$ is a time-dependent variable whose evolution is described by

$$\frac{dx(t)}{dt} = \epsilon x(t), \tag{2.20}$$

where ϵ is real valued and is equivalent to Eq. (2.6) with the initial condition $x(0) = x_0$. We have introduced the symbol ϵ to describe the situation where it is "small," but we now allow its sign to change. This equation, of course, can be integrated directly. In our previous discussion, this quantity would generally have been negative-valued, imposing a solution that decays exponentially. However, if ϵ were to change sign, exponential decay in the amplitude would be replaced by exponential growth $\exp(\epsilon t)$. The only characteristic of the solution that remains unchanged is that the sign of the variable $x(t)$ does not change. We observe that ϵ also describes the eigenvalue associated with the differential operator, in this case, the factor that appears inside an exponential. This presents a *bifurcation*, often designated a *Hopf bifurcation* for nonlinear problems in honor of one of the founders of nonlinear dynamics, and presents a qualitative change in behavior. If $x(t)$ described, for example, a small perturbation in a physical system, the emergence of a positive-valued ϵ would imply that the perturbation would grow without limit at an exponential pace. This is the essence of chaos: Any system possessing the potential for exponential growth in perturbations is fundamentally unstable. Lorenz (1963b) considered the limitations of weather forecasting and remarked that "one flap of a sea gull's wings would be enough to alter the course of the weather forever."

2.2 Second-Order Ordinary Differential Equations

In geophysics, second-order ordinary differential equations have a pivotal role. We will begin by revisiting Newton's second law, which exposes many features attendant to nonlinear problems while remaining exactly soluble—albeit in one spatial dimension. We will then proceed to explore second-order linear homogeneous ordinary differential equations with constant coefficients, in order to identify the associated solutions, and go on to introduce the role of forcing, that is, the inhomogeneous term. This will serve as a springboard for other topics emergent from second-order ordinary differential equations and approximate methods of solution.

It is noteworthy, however, that a set of coupled first-order ordinary differential equations can generally be converted into a single higher-order ordinary differential equation. We will say more about higher-order equations in due course, but a physically motivated example is in order here. Consider the coupled set of equations, taken from classical mechanics, for one-dimensional particle motion under the influence of a potential $V(x)$, namely,

$$m\dot{x} = p$$
$$\dot{p} = -\frac{dV(x)}{dx}, \tag{2.21}$$

where x is the particle's position, m is its mass, and p is its (linear) momentum. We recognize this as Newton's second law, and (2.21) can be written

$$m\ddot{x} + \frac{dV(x)}{dx} = 0 \qquad (2.22)$$

which is a second-order, first-degree ordinary differential equation. Importantly, when we multiply it by $\dot{x}$, we observe that

$$m\dot{x}\ddot{x} + \dot{x}\frac{dV}{dx} = \frac{d}{dt}[\tfrac{1}{2}m\dot{x}^2 + V(x)] = 0, \qquad (2.23)$$

from which we deduce the *conservation law*

$$\tfrac{1}{2}m\dot{x}^2 + V(x) = \frac{1}{2m}p^2 + V(x) = E, \qquad (2.24)$$

where E is the energy of motion. We have, at the initial time t_0, an initial position x_0 and initial velocity v_0, which is $\dot{x}$ at that time. E is also referred to as a *constant of motion* and is related to the initial conditions according to

$$E = \tfrac{1}{2}mv_0^2 + V(x_0). \qquad (2.25)$$

With this outcome, we convert the complete solution for this problem into a *quadrature* by noting that

$$\dot{x} = \pm\sqrt{\frac{2[E - V(x)]}{m}}, \qquad (2.26)$$

according to which we can write

$$\pm\int_{x_0}^{x}\sqrt{\frac{m}{2[E - V(x')]}}\,dx' = t - t_0, \qquad (2.27)$$

where the sign is established according to the direction of motion in the vicinity of the *turning point*, that is, the location where the potential and total energies are equal, and the kinetic energy vanishes. Since the complete solution can be expressed in this way, we call it *integrable*. These are elementary properties of *Hamiltonian* systems, which provide an important theoretical framework for solving a broad class of problems in analytical mechanics. In many instances, we can identify the energy E in the system with the so-called *Hamiltonian*.

We present here two specific examples to illustrate some of the features of this problem and the geometry that emerges. Consider the *simple harmonic oscillator*

$$m\ddot{x} + kx = 0, \qquad (2.28)$$

where m is the mass, k is the spring constant, and $\omega = \sqrt{k/m}$ is the frequency of the oscillator. Suppose that ℓ is a typical measure of the spring displacement. We now define dimensionless length x' and time t' variables according to

$$x' = x/\ell \quad \text{and} \quad t' = \omega t. \tag{2.29}$$

The dimensionless form of the force equation becomes

$$\frac{\mathrm{d}^2 x}{\mathrm{d}t^2} + x = 0, \tag{2.30}$$

where we have dropped the primes $'$ for convenience. We now employ $p = \dot{x}$ in our dimensionless system, and the energy conservation law becomes

$$\tfrac{1}{2}p^2 + \tfrac{1}{2}x^2 = E \tag{2.31}$$

where E is our scaled energy. The method of *phase trajectories* provides a helpful way to visualize some aspects of this problem. This involves plotting $x(t)$ and $p(t)$ where the time t is implicit, and we refer to the ensuing curve as a *phase portrait* (Figure 2.1).

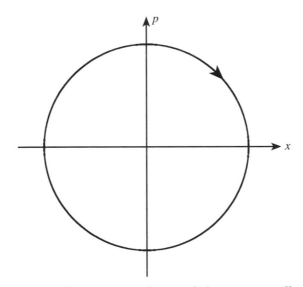

Figure 2.1.   Phase portrait for simple harmonic oscillator.

We observe that the phase trajectory is a circle of radius $\sqrt{2E}$, where the trajectory is clockwise. The extreme values of x achieved are the turning points mentioned previously and correspond to the positions where the trajectory, which is evidently clockwise, reverses its course. Intuitively, this figure also describes the cycle of energy conversion from potential to kinetic and back.

Finally, it is useful to consider how (2.31) would appear if we regarded $E(x, p)$ as a function and produced a three-dimensional representation of this figure, namely, a paraboloid. The specific two-dimensional plot that we have can be considered to be the outcome of taking a slice through that parabola, which we then project onto the x-p plane. Similarly, we can produce a set of "contour plots" for this geometry that have come to be called *level surfaces*, a topic we will revisit as we explore more complicated dynamics.

2.2.1 Linear Second-Order Differential Equations

Linear second-order ordinary differential equations occupy a pivotal role in geophysics, and we will focus here on their solution before going on to address nonlinear problems. Consider, therefore, the general equation

$$a(t)\ddot{x}(t) + b(t)\dot{x}(t) + c(t)x(t) = f(t); \qquad (2.32)$$

we will defer our description of the meaning and properties of the possibly time-dependent coefficients a to c.

In the previous section, we have observed that linear first-order ordinary differential equations have exponentials associated with their solution. Second-order linear ordinary differential equations generally have two solutions. The homogeneous solution to a linear second-order ordinary differential equation is the linear superposition of the two linearly independent functions. A useful way to think of this is that the exponentials that ultimately characterize second-order equations have solutions determined from the solution of a quadratic equation. A simple case in point emerges from the equation

$$\frac{\mathrm{d}^2 x(t)}{\mathrm{d}t^2} = x(t); \qquad (2.33)$$

if we assume that solutions to this equation have the form $\exp(\kappa t)$, we find that $\kappa = \pm 1$ so that a general solution to this equation is

$$x(t) = a_+ \exp(+t) + a_- \exp(-t). \qquad (2.34)$$

The constants $a_\pm$ are determined to match initial and/or boundary conditions. (The existence and uniqueness of solutions is only guaranteed for initial value problems, not boundary value problems.) It is clear, if we are dealing with some phenomena propagating from the origin to very large values of t, that $a_+ = 0$ so that the solution does not diverge and that $a_- = x(0)$ to match the initial conditions. In this instance, we see that the damping solution is physically relevant, and is referred to as the *regular solution*, while the growing solution is nonphysical, and is referred to as the *irregular solution*. A similar situation emerges if we were to go back in time to very large negative values of t. For time-dependent $a(t)$, $b(t)$, and so on, the situation is more complicated, particularly when any of these coefficients vanish or become singular at finite time, and the need to establish which solution is regular and which is irregular becomes of paramount importance. The need to identify which solution applies to a given physical situation is a common feature of second-order equations. This also identifies a problem that emerges in the computational solution of ordinary differential equations: Numerical error due to approximations in the method or due to rounding errors will introduce into the computed solution a component of the irregular solution that, ultimately, will overwhelm the regular solution rending the numerical results invalid. We will briefly address this issue in our last chapter.

2.2.2   Green's Functions

We will approach this in a series of graduated steps, beginning with the case $a(t) = 1$ and $b(t) = c(t) = 0$. We recognize this as being the time-dependent form of Newton's second law, with no potential energy term, subject to a force (per unit mass) $f(t)$, namely,

$$\ddot{x}(t) = f(t), \tag{2.35}$$

which has homogeneous solutions 1 and t; we recognize that a constant happens to be a function that does not vary. The initial conditions for $t = 0$ are $x(0) = x_0$ and $\dot{x}(0) = v_0$, as before. Hence, a linear combination of the homogeneous solutions can be expressed as $d_1 + d_2 t$, where d_1 and d_2 are constants. A general solution for $t \geqslant 0$ can be obtained by integrating (2.35) twice, namely,

$$\dot{x}(t) = \int_0^t dt'' f(t'') + v_0$$

$$x(t) = \int_0^t dt' \int_0^{t'} dt'' f(t'') + v_0 t + x_0. \qquad (2.36)$$

We reorganize the integration to get

$$x(t) = \int_0^t dt'' f(t'') \int_{t''}^t dt' + x_0 + v_0 t$$

$$= \int_0^t dt'' f(t'')(t - t'') + x_0 + v_0 t. \qquad (2.37)$$

We recognize the first term on the right-hand side as being a particular solution while the two additional terms, selected to match the initial conditions, present a linear combination of the homogeneous solutions. However, we note that the term $t - t''$ is the ramp function $R(t - t'')$ that we introduced in the previous chapter. Accordingly, we can now write

$$x(t) = \int_0^\infty dt'' f(t'') R(t - t'') + x_0 + v_0 t, \qquad (2.38)$$

where the limits of integration have now been extended consistent with the definition of the ramp function. Recalling (1.82) that the second derivative of the ramp function is the Dirac δ function, we instantly recover our original equation (2.35).

We can now employ this result to introduce the concept of a *Green's function*. Simply stated, a Green's function is the kernel of an integral that provides a solution to an inhomogeneous, linear ordinary or partial differential equation in the form of an integral over the Green's function multiplied by the inhomogeneous source term. This methodology is one of the great achievements of mathematical physics, and many more details can be found in the textbooks by Stakgold (1998), Morse and Feshbach (1999), Courant and Hilbert (1962), Mathews and Walker (1970), Butkov (1968), as well as in chapter 4 here. The fundamental idea here is that the Green's function, say $G(t, t')$, is one that, when acted upon by the homogeneous operator at hand, can be expressed as a Dirac δ function. In the example above, the operator $\mathcal{L}$ is simply

$$\mathcal{L} = \frac{d^2}{dt^2} \qquad (2.39)$$

and the ramp function $R(t - t')$ is the associated Green's function, that is,

$$G(t, t') = R(t - t'). \tag{2.40}$$

A useful way to think of a Green's function is that it provides the response to an impulse in an inhomogeneous ordinary or partial differential equation defined on a specified domain with specified initial and/or boundary conditions. Green's functions exploit the superposition principle, the notion that a convolution of a special "benchmark" solution with the inhomogeneous source provides the complete solution to the problem. We will utilize these properties elsewhere in this chapter and in chapter 4.

As the next step in our discussion, let us discuss Helmholtz's equation in one space dimension, which will be introduced in chapter 4, for $0 \leqslant x \leqslant L$,

$$\frac{d^2 y(x)}{dx^2} + k^2 y(x) = 0, \tag{2.41}$$

subject to boundary values $y(0) = y(L) = 0$, sometimes referred to as "clamped." This problem is the simplest example of a *boundary value problem*, in contrast with an *initial value problem*, as it is necessary for the solution to satisfy what might otherwise be regarded as initial value constraints at the two ends of the domain. Since this equation must be satisfied everywhere inside this domain and expecting to find solutions of the form $\exp(i\kappa x)$, we insert this trial solution into the preceding equation and obtain

$$\kappa^2 = k^2, \tag{2.42}$$

showing that the homogeneous solutions $y_\pm(x)$ for this problem are given by

$$y_\pm(x) = \exp(\pm ikx). \tag{2.43}$$

Hence, the solution for Helmholtz's equation must appear as a linear combination of these having the form $d_+ y_+(x) + d_- y_-(x)$, that is,

$$y(x) = d_+ y_+(x) + d_- y_-(x) = d_+ \exp(+ikx) + d_- \exp(-ikx). \tag{2.44}$$

In order to satisfy the left boundary condition, we observe that $d_- = -d_+$ and, therefore, identify the consistency condition

$\exp(i\kappa L) = \exp(-i\kappa L)$. Hence, it follows that

$$\exp(2i\kappa L) = 1 \qquad (2.45)$$

and we observe that κ must satisfy the *eigenvalue* condition

$$2\kappa L = 2n\pi \qquad (2.46)$$

for integer-valued n, that is,

$$\kappa = \frac{n\pi}{L}, \qquad (2.47)$$

and the *eigenmode* for what is a standing wave has the form

$$y(x) = A \sin\left(\frac{n\pi x}{L}\right), \qquad (2.48)$$

where A is the wave amplitude.

We return now to the general equation (2.32) over the same domain where we assume that a, b, and c are constant coefficients. Seeking homogeneous solutions of the form $\exp(\mu x)$, we observe that μ must satisfy the quadratic equation

$$a\mu^2 + b\mu + c = 0, \qquad (2.49)$$

so that

$$\mu_{\pm} = \frac{-b \pm \sqrt{b^2 - 4ac}}{2a}. \qquad (2.50)$$

If the same left-hand boundary value applies, it again follows that the homogeneous solution emerges with the form

$$y(x) = A[\exp(\mu_+ x) - \exp(\mu_- x)]. \qquad (2.51)$$

In the case of the boundary value problem, the eigenvalue described in Eq. (2.46) must satisfy the condition

$$n = \frac{L\sqrt{4ac - b^2}}{2\pi a}. \qquad (2.52)$$

Importantly, it follows that the quantity inside the square root must be positive for a solution to the boundary value problem to exist. This is an example of a *Sturm–Liouville* eigenvalue problem, which we will discuss later in this chapter.

2.2.3 LRC Circuits and Visco-Elastic Solids

In order to explore the inhomogeneous form of (2.32), let us return to the time domain and consider a driven LRC electrical circuit. This example is especially important in geophysics, as such circuits have been regarded as useful analogues for geodynamics and the Earth's dynamo. Figure 2.2 provides, in its two panels, an illustration of the LRC model and what has come to be called the *Maxwell visco-elastic* model (Stacey and Davis, 2008). Situations where the spring and viscous damping components are in parallel are referred to as *Kelvin–Voigt* models and are sometimes encountered in geophysical and materials science applications as analogues of LRC circuits (Garland, 1979; Stacey and Davis, 2008; Turcotte et al., 2002; Bullen, 1963).

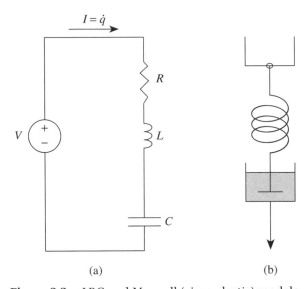

$$I = \dot{q}$$

(a)   (b)

Figure 2.2. LRC and Maxwell (visco-elastic) models.

Accordingly, we now write for a simple LRC circuit the governing differential equation for the current $I(t) = \dot{q}(t)$, where q is the charge and I now represents the current $\dot{q}$, namely,

$$L\ddot{q}(t) + R\dot{q}(t) + \frac{1}{C}q(t) = V(t), \qquad (2.53)$$

where $V(t)$ is the applied voltage. This is related to the second-order ordinary differential equation (2.32) considered earlier and contains damping. Panel (a) in Figure 2.2 describes the LRC circuit while panel (b) describes a Maxwell visco-elastic solid. The spring in the latter describes the harmonic oscillator term, corresponding to q/C in (2.53), the viscous medium introduces a resistive force akin to $R\dot{q}$ in the former equation, and the weight's acceleration is analogous to the $L\ddot{q}$ term.

We begin by exploring the homogeneous solutions of the form $\exp(i\omega t)$ to Eq. (2.53) and observe that

$$\omega_\pm = i\frac{R}{2L} \pm \frac{\sqrt{4L/C - R^2}}{2L}. \tag{2.54}$$

In the general case with no restrictions on the sign of $4L/C - R^2$, there can be underdamped, critically damped, and overdamped cases depending on the sign of the discriminant: A negative discriminant is overdamped, a vanishing one is critically damped, and a positive discriminant is underdamped. Assuming that $L/C \gg R^2$ (underdamped), as is usually the case, it follows that

$$\omega_\pm \approx \pm\sqrt{\frac{1}{LC}} + i\frac{R}{2L}. \tag{2.55}$$

This system is fundamentally oscillatory and has a frequency of $\sqrt{1/LC}$. Owing to the presence of the resistor in the LRC case or viscosity in the visco-elastic case, we observe that the energy (or the square of the amplitude) in this system decays exponentially as $\exp(-Rt/L)$. It is common in geophysics (see, e.g., Stacey and Davis (2008)) to define the Q of the system according to the expression

$$\frac{2\pi}{Q} = -\frac{\Delta E}{E}, \tag{2.56}$$

where ΔE is the energy drop during one oscillation period $\tau \approx 2\pi/\omega$, which gives

$$Q \approx \frac{1}{R}\sqrt{\frac{L}{C}}. \tag{2.57}$$

This characterization of dissipative problems is very commonly employed in geophysics and other disciplines.

2.2.4 Driven Oscillators, Resonance, and Variation of Constants

We shall now explore the behavior of the driven simple harmonic oscillator by adding the "spring term" to Eq. (2.35) so that it becomes the inhomogeneous simple harmonic oscillator equation

$$\ddot{x}(t) + \omega^2 x(t) = f(t). \tag{2.58}$$

In many ways, this may be regarded as a good mathematical model for a seismometer. We can factor the left-hand side of this expression to obtain the associated integrating factors:

$$\ddot{x} + \omega^2 x = \left(\frac{d}{dt} + i\omega\right)\left(\frac{d}{dt} - i\omega\right)x$$
$$= \exp(-i\omega t)\frac{d}{dt}\left[\exp(2i\omega t)\frac{d}{dt}(\exp(-i\omega t)x)\right].$$

$$(2.59)$$

We equate the latter expression with $f(t)$; we can now integrate this expression in two steps, just as we solved for a particular solution of (2.35) by repeated integration. We obtain a particular solution x_{part}, given by

$$x_{\text{part}}(t) = \int_0^\infty dt' f(t')\frac{\sin[\omega(t - t')]}{\omega}H(t - t'),\qquad(2.60)$$

to which we add a linear combination of the homogeneous solutions $\exp(\pm i\omega t)$ to obtain the general or complete solution. Here, $H(t)$ is the Heaviside or step function. Further, we observe that the integrand terms to the right of the driver $f(t')$, namely,

$$G(t, t') = \frac{\sin[\omega(t - t')]}{\omega}H(t - t'),\qquad(2.61)$$

constitute the Green's function $G(t, t')$ for the driven equation.

We will now provide another methodology, due originally to Euler and to Lagrange—see Bender and Orszag (1999) and Newman and Efroimsky (2003) for more details. Let us assume that a general solution to (2.58) can be expressed using a linear combination of the homogeneous solutions, where we assume that the linear combination coefficients $d_\pm$ are now functions of time, namely,

$$x(t) = d_+(t)\exp(+i\omega t) + d_-(t)\exp(-i\omega t),\qquad(2.62)$$

where the properties of $d_\pm(t)$ must be established. This approach, for obvious reasons, is often called the method of *variation of constants* or the method of *variation of parameters*. It is noteworthy that this approach is also valid for general first-order ordinary differential equations, as we observed earlier. We now take the derivative of this expression but assume that the contributions emerging from $\dot{d}_\pm()$ vanish, that is,

$$0 = \dot{d}_+(t)\exp(+i\omega t) + \dot{d}_-(t)\exp(-i\omega t);\qquad(2.63)$$

the underlying rationale here is that the "constants" or "parameters" do not in some sense vary—and the latter expression provides a definition for what is meant by that. [The methodology due to Euler and to Lagrange arbitrarily established a constraint on the contribution to the velocity arising from the time derivatives of the "constants." Newman and Efroimsky (2003) show that we are at liberty to choose essentially any constraint for the "constants" and yet obtain the same result. This feature of differential equations is called *gauge freedom*.] Accordingly, it follows that

$$\dot{x}(t) = d_+(t)i\omega\exp(+i\omega t) - d_-(t)i\omega\exp(-i\omega t). \quad (2.64)$$

We take the time derivative of this expression and make use of (2.58) to obtain

$$f(t) = \dot{d}_+(t)i\omega\exp(+i\omega t) - \dot{d}_-(t)i\omega\exp(-i\omega t). \quad (2.65)$$

This latter equation together with (2.63) present two equations in the time derivative of the unknown quantities $d_+(t)$ and $d_-(t)$, namely,

$$\dot{d}_+(t) = +\frac{f(t)}{2i\omega}\exp(-i\omega t)$$

$$\dot{d}_-(t) = -\frac{f(t)}{2i\omega}\exp(+i\omega t). \quad (2.66)$$

Solving these equations, together with the initial conditions, recovers Eq. (2.60). Importantly, the method of variation of constants eliminates the need to be able to identify and utilize the underlying integrating factors, although it does require the identification of the homogeneous solutions.

To complete this section, there is an additional problem that merits attention owing to its prominence in geophysics—the role of *resonance*. This phenomenon is particularly prominent in planetary physics applications (Murray and Dermott, 1999; De Pater and Lissauer, 2010). For example, the period or *mean motion* of Saturn is approximately 30 years while that of Jupiter is 12, thereby producing a nearly 5:2 resonance and was identified by Laplace. A more accurate calculation of the ratio reveals a mean motion resonance of approximately 900 years, which is commonly referred to as the *great inequality*. Resonant effects between the giant planets and asteroids give rise to the *Kirkwood gaps* in the asteroid belt, while resonant effects between

Saturn's satellites and its rings produce the *Cassini divisions*. Consider now the equation

$$\ddot{x}(t) + \omega^2 x(t) = \mathcal{A}\cos(\Omega t), \tag{2.67}$$

where $\omega \neq \pm\Omega$. Since $\sin(\Omega t)$ is an eigenmode of the operator d^2/dt^2, that is, the outcome of that operation is proportional to the original quantity, we consider a trial solution $A\cos(\Omega t)$. We observe that

$$A = \frac{\mathcal{A}}{\omega^2 - \Omega^2} = \frac{\mathcal{A}}{2\omega}\left[\frac{1}{\omega - \Omega} + \frac{1}{\omega + \Omega}\right]. \tag{2.68}$$

(There are applications where the sign of ω and Ω may not be positive, so we have been careful to identify the possible cases that emerge.) We observe, as the frequencies ω and Ω become closer together, that the amplitude of the response will grow and, ultimately, could become singular. We need to explore the limit $\Omega \to \omega$: Using the methods already established we can find the resonant solution that applies. In particular, we observe and can readily verify that a particular solution for this problem is

$$x(t) = \frac{\mathcal{A}t}{2\omega}\sin(\omega t). \tag{2.69}$$

Note that the amplitude of the sinusoidal response is now modulated by the time t. We refer to this kind of behavior as presenting a *secular solution* inasmuch as the sinusoid now resides inside an envelope that grows linearly in time. Finally, we observe that linear problems only support 1-to-1 resonances where the frequency of the driver and the natural response frequency of the system are the same. Richer resonant phenomena emerge in nonlinear problems. For example, if the forcing function had the form $\mathcal{A}\cos^3(\Omega t)$, the triple angle identity (1.73) would result in our being able to write

$$f(t) = \mathcal{A}[\tfrac{1}{4}\cos(3\Omega t) + \tfrac{3}{4}\cos(\Omega t)]. \tag{2.70}$$

Thus, we expect that resonance and secular behavior will emerge if ω is close to Ω or to 3Ω. In complex nonlinear problems, especially those emergent in perturbation theory, resonances occur for rational Ω/ω, that is, when the ratio is close to some m/n, where m and n are integers, especially if they are small. This, for example, is an important factor in the evolution of orbits in our solar system, including the so-called *Kirkwood gaps* in the asteroid belt and the *Cassini divisions* in Saturn's rings.

Before moving on to nonlinear problems, we will now address an important class of second-order differential equations encountered in geophysics, among other fields, and an approximate yet remarkably effective methodology for evaluating the solution.

2.2.5 JWKB Method, Riccati Equation, and Adiabatic Invariants

Our focus in much of this chapter has been upon linear second-order ordinary differential equations with constant coefficients. However, many problems emerge where the term in $x(t)$ is itself a function of time; we will employ

$$\ddot{x}(t) + \omega^2(t)x(t) = 0 \tag{2.71}$$

as a prototype, where ω is now regarded as a slowly varying function of time. Much of the literature on this problem has emerged owing to its equivalence to *Schrödinger's equation* (Morse and Feshbach, 1999) in quantum mechanics, where x is replaced by the wave function and the dependent variable is the spatial coordinate. Such problems often emerge in the context of eigenvalue determinations, just as we observed in the case of Helmholtz's equation, and substantial effort must be invested in insuring that the original equation has *Sturm–Liouville form* (Morse and Feshbach, 1999)—that is, effectively

$$\frac{\mathrm{d}}{\mathrm{d}t}\left[p(t)\frac{\mathrm{d}x(t)}{\mathrm{d}t}\right] + [q(t) + \lambda r(t)]q(t)x(t) = 0,$$

where λ is an eigenvalue—so that certain "variational" properties are preserved. We will not pursue quantum mechanical applications here. The mathematical geophysicist Jeffreys (1925) had a prominent role in early studies of this problem, and Wentzel, Kramers, and Brillouin (Kemble, 2005) expanded on the methodology during the early history of quantum mechanics, seeking to bridge the methodology for solving quantum mechanical problems with ideas emergent from classical physics. Kemble (2005) is an especially helpful source for addressing issues associated with classical turning points. This methodology should be referred to as the JWKB method, but the contributions by Jeffreys have been overlooked and the solution is referred to as the WKB method. Bender and Orszag (1999) provide an excellent overview of the methodology. Newman and Thorson (1972a,b) developed an iterative class of methods for improving the accuracy of the solution.

In our treatment of the forced simple harmonic oscillator (2.58), we assumed that the spring frequency was a constant in seeking the homogeneous solutions $\exp(\pm i\omega t)$. It is natural to expect, therefore, that the homogeneous solutions might be related to the continuously varying version of the above, namely,

$$x_{\pm}(t) = \exp\left[\pm i\int^{t}\omega(t')\,dt'\right]. \tag{2.72}$$

If we introduce this expression into Eq. (2.71), we observe that there is an error of order $\dot{\omega}(t)$ relative to the $\omega^2(t)$ term. Specifically comparing these two terms, we observe that their ratio $\mathcal{R}$ characterizes the local oscillatory time relative to the e-folding time for change in the oscillatory frequency, that is, $\mathcal{R} = |\dot{\omega}(t)/\omega^2(t)|$. Accordingly, let us use this insight to find an exact solution to (2.71):

$$x(t) = \exp\left[i\int^{t}W(t')\right]dt', \tag{2.73}$$

whereupon we observe that the quantity $W(t)$ satisfies the first-order equation that is quadratic in the unknown (sometimes referred to as a *Riccati equation*)

$$\frac{dW(t)}{dt} = i\omega^2(t) - iW^2(t), \tag{2.74}$$

which is the form of an ordinary differential equation that we would normally attempt to solve but will rewrite instead as

$$W^2(t) = \omega^2(t) + i\frac{dW(t)}{dt}. \tag{2.75}$$

In the latter expression, $W(t) \approx \pm\omega(t)$ is a good approximation so long as $\mathcal{R}$ is very small. (Note that this problem could be approached by using a formal perturbation expansion, such as that presented later in this chapter. However, the approach presented here provides a simpler way to obtain the usual JWKB approximation.) Hence, we will replace the trailing term by $\pm i\,d\omega(t)/dt$ and obtain

$$W_{\pm}^2(t) \approx \omega^2(t) \pm i\frac{d\omega(t)}{dt}. \tag{2.76}$$

Taking the square root of both sides, but preserving the $\pm$ sense, we obtain

$$W_{\pm}(t) = \pm\omega(t)\left[1 \pm \frac{i}{2\omega^2(t)}\frac{d\omega(t)}{dt}\right]. \qquad (2.77)$$

We observe that the correction to $W(t)$ is of order $\mathcal{R}$, as expected, but seek to introduce this approximate solution into (2.73) by

$$\begin{aligned} W_{\pm}(t) &= \exp\left[\pm i\int^t \omega(t')\,dt'\right]\exp\left[\int^t -\frac{1}{2\omega(t')}\frac{d\omega(t')}{dt'}dt'\right] \\ &= \frac{1}{\sqrt{\omega(t)}}\exp\left[\pm i\int^t \omega(t')\,dt'\right], \qquad (2.78) \end{aligned}$$

having observed that the trailing exponential in the first line is directly integrable. We can also introduce real-valued forms of the solution by using cosines and sines, namely,

$$\begin{aligned} W_c(t) &= \frac{1}{\sqrt{\omega(t)}}\cos\left[\int^t \omega(t')\,dt'\right] \\ W_s(t) &= \frac{1}{\sqrt{\omega(t)}}\sin\left[\int^t \omega(t')\,dt'\right], \qquad (2.79) \end{aligned}$$

and linear combinations of these real solutions can readily be employed. While we immediately recognize that the exponential/sinusoidal term describes the local oscillations, the amplitude term is less obvious. (In quantum mechanical applications, it can be shown to be related to the uncertainty principle.) We will now show why the leading term, for our slowly time-varying simple harmonic oscillator is to be expected.

Let us return to Eq. (2.71), which we multiply by by $x(t)$, namely,

$$\begin{aligned} 0 &= x(t)\ddot{x}(t) + \omega^2(t)x^2(t) \\ &= \frac{d}{dt}[x(t)\dot{x}(t)] - \dot{x}^2(t) + \omega^2(t)x^2(t). \qquad (2.80) \end{aligned}$$

This is sometimes called the *virial* (Goldstein et al., 2002). As before, we now identify the "energy" $\mathcal{E}$ by

$$\mathcal{E} = \tfrac{1}{2}\dot{x}^2(t) + \tfrac{1}{2}\omega^2(t)x^2(t). \qquad (2.81)$$

However, using our solutions (2.79), we observe that $\mathcal{E}$ is no longer a constant but the quantity

$$\mu = \frac{\mathcal{E}(t)}{\omega(t)} + \mathcal{O}(\mathcal{R}) \qquad (2.82)$$

is approximately constant and is referred to as the (first) *adiabatic invariant*. Furthermore, this expression is not exact but has a relative error $\mathcal{O}(\mathcal{R})$.

There are many circumstances in geophysics where adiabatic invariants play an important role. For example, in charged particle trajectories along geomagnetic field lines, the associated magnetic field will change. The simple harmonic oscillator emerges from the Lorentz force law for particle motion in a magnetic field, and the gyrofrequency or Larmor frequency corresponds to the spring frequency that we have employed here. The energy that we have calculated corresponds to that of the gyromotion of particles around field lines and provides a remarkably accurate diagnostic in observations of energetic particle flows.

Since we have already introduced the Riccati equation, let us explore one additional example relevant to geophysics. Suppose that we are investigating the velocity $v(t)$ of an object falling under the acceleration of gravity, g, and resisted by the atmosphere. See, for example, Feynman et al. (1989) for a physical description of atmospheric friction and Davis (1960) for a mathematical treatment of its effects. The resistive drag experienced by the object varies as $\alpha v^2(t)$, where α is proportional to the atmospheric density and incorporates the geometry of the object, a result first established by Lord Rayleigh. Thus, our dynamical equation has the form

$$\dot{v} = g - \alpha v^2. \tag{2.83}$$

The terminal velocity v_t corresponds to a balance between gravitational acceleration and frictional drag, namely,

$$v_t = \sqrt{\frac{g}{\alpha}}, \tag{2.84}$$

and is called a *stationary solution* since it corresponds to a solution that does not change over time. Let us look for a solution of the form

$$v = v_t + \frac{1}{u}, \tag{2.85}$$

whereupon we observe that

$$\dot{u} = 2\alpha v_t u + \alpha. \tag{2.86}$$

We immediately recognize that this equation has the same form as (2.13) with constant coefficients, which we can solve explicitly.

Accordingly, we see that u will grow exponentially in time, corresponding to v approaching terminal velocity at an exponential rate. For first-order ordinary differential equations having quadratic nonlinearity including variable coefficients, a generalization of (2.83), the Riccati equation can also be transformed into a second-order linear ordinary differential equation whose solution we have already discussed. We now wish to explore richer forms of nonlinearity.

2.2.6 Nonlinearity and Perturbation Theory

Let us now consider a more complicated yet familiar dynamical problem, a pendulum of length ℓ and mass m undergoing gravitational acceleration g. Newton's second law for the (dimensionless) angle θ follows immediately as

$$m\ell\ddot{\theta} + mg\sin(\theta) = 0. \tag{2.87}$$

We introduce a frequency-like constant quantity $\omega \equiv \sqrt{g/\ell}$, a dimensionless time $t' = \omega t$, replace θ by x, and obtain

$$\ddot{x}(t) = -\sin[x(t)]. \tag{2.88}$$

Upon multiplying by $\dot{x}$, which we will later replace as before by p, we obtain the conservation law

$$\tfrac{1}{2}p^2 + [1 - \cos(x)] = E. \tag{2.89}$$

Here, we have employed prevailing custom in adding a constant to the left-hand side to assure that it is always positive. We now superpose phase portraits for the pendulum problem, labeling each individual phase portrait by the dimensionless value of E that pertains.

This has the appearance of a set of contours that are sometimes referred to as *level surfaces* of the Hamiltonian, as we noted earlier, which is the energy quantity that we have calculated, if one regarded the energy (2.89) as a function of x and p whose contours we plot in Figure 2.3. Similarly, we can envisage the geometry of a three-dimensional rendering. As in the case of the simple harmonic oscillator while in the vicinity of the origin, we are dealing with a surface that is effectively paraboloidal. However, as we move away from the origin, especially toward x-values of $\pm\pi$, the surface that emerges is readily

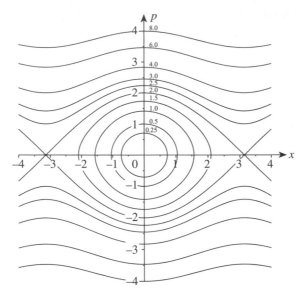

Figure 2.3. Phase portrait for a pendulum.

observed to be a *saddle point*. To make this imagery clearer, we show in Figure 2.4 the saddle point we would associate with $\frac{1}{2}p^2 - \frac{1}{2}x^2 + 2 = E$, which approximately describes the pendulum problem in the vicinity of $(\pi, 0)$, that is, we have shifted x by π.

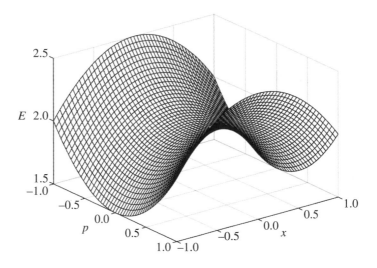

Figure 2.4. Saddle point.

Indeed, some unusual geometrical features have appeared.

In the vicinity of the origin, the contours appear approximately to be concentric circles, which is consistent with (2.89) for small angles x of pendulum motion, where the trajectories are clockwise. As E approaches 2, we encounter a situation where the pendulum is just able to swing into a vertical position and the angle of oscillation approaches $\pm\pi$ radians at infinite time. For reasons that will become evident later, we refer to the point $(0, 0)$ as a *center* and the points $(\pm\pi, 0)$ as saddle points. The contour connecting the two saddle points (as well as others situated at odd integer multiples of π) is referred to as a *separatrix*. Trajectories that lie inside the separatrices are said to describe *libration*, while those that lie outside describe *rotation*, inasmuch as the pendulum will continue going around and around. Having observed how the geometry of the solution appears, for small amplitudes, very similar to that of the simple harmonic oscillator, we now wish to introduce some ideas emerging from perturbation theory—see, for example, Nayfeh and Balachandran (1995), Bender and Orszag (1999), and Strogatz (1994) for more elaborate treatments.

Let us begin with Eq. (2.88), expand the sine function keeping a few terms, and employ an expansion parameter ε. The idea here is to establish a differential equation that corresponds to the simple harmonic oscillator when $\varepsilon = 0$, and to the pendulum problem when $\varepsilon = 1$. (While our focus here is on ordinary differential equations, a similar procedure is employed in applications to partial differential equations.) We now explore the behavior of the differential equation assuming that ε is "small":

$$\ddot{x}(t) = -x(t) + \varepsilon\{x(t) - \sin[x(t)]\}$$

$$= -x(t) + \varepsilon\left\{\frac{x^3(t)}{3!} + \cdots\right\}. \qquad (2.90)$$

We will employ a perturbation expansion in terms of the parameter ε of the form

$$x(t) = x^{(0)}(t) + \varepsilon x^{(1)}(t) + \cdots. \qquad (2.91)$$

Intuitively, we select the zero-order term to correspond to the amplitude in the absence of nonlinearity, and introduce additional terms to correspond to the perturbation that emerges. (Commonly, perturbation expansions are only performed to first-order, although using computer algebra engines like *Maple* and *Mathematica*, it is possible to explore higher-order expansions.) Matching increasing powers in the parameter ε, we obtain

$$\ddot{x}^{(0)}(t) = -x^{(0)}(t)$$

$$\ddot{x}^{(1)}(t) = -x^{(1)}(t) + \frac{x^{(0)^3}(t)}{6}. \tag{2.92}$$

Suppose we express our solution for $x^{(0)}(t)$ as $A\cos(t - t_0)$ where, without loss of generality, t_0 was selected to that $\dot{x}(t_0) = 0$ and $x(t_0) = A$. The homogeneous solutions for $x^{(1)}(t)$ are the same as for $x^{(0)}(t)$, but we must now introduce the inhomogeneous term emerging from the cubic term in $x^{(0)}(t)$.

We immediately observe that a resonant term is present in the second equation, from the triple-angle formula, and that its evolution now includes a term that is linear in time, often referred to as *secular growth*. Higher powers of time emerge as additional secular growth terms in the higher-order perturbations as well. We observe here a deficiency in many perturbation methods: They can produce growth terms that ultimately will make the underlying conservation laws appear to fail. The important point here is that perturbation methods can describe the initial departure from the unperturbed solution, but a more detailed analysis is essential in obtaining a deeper understanding of the solution.

It should be clear, in looking at the phase portrait for the pendulum, that the perturbation terms including powers in time are describing the departure from the otherwise circular trajectories. However, as the amplitudes of the different orders of perturbation grow large, higher powers of time become necessary to curtail the growth. Indeed, there are numerous instances where the higher-order perturbations presented the cumulative effect of representing a trigonometric function that, in essence, was attempting to change the underlying frequency of oscillation to a much lower one. To understand this, consider the approximate differential equation

$$\ddot{x}(t) = -x(t)\left[1 - \frac{x^2(t)}{6}\right]. \tag{2.93}$$

Multiplying this by $\dot{x}(x)$, we observe that the quantity

$$\mathcal{E} = \frac{\dot{x}^2(t)}{2} + \frac{x^2(t)}{2} - \frac{x^4(t)}{24} \tag{2.94}$$

is conserved. We note how the last term tends to flatten the emergent circular trajectories, but how, one might ask, is the frequency of the pendulum altered? To answer this question, it is useful to construct a variant of this problem since it is clear that the bracketed quantity in (2.93) is necessarily smaller than 1 in magnitude and, thereby, will reduce the frequency. To estimate its influence, let us replace the $x^2(t)$ term by its time average $\langle x^2(t) \rangle$; we can employ for this purpose the unperturbed solution to the original ordinary differential equation, whereupon we obtain an estimate of $A^2/2$ for that average and a reduction in the frequency (squared) to $1 - A^2/12$. The presence of the conservation law, together with an appreciation for the frequency reduction, rounds out our understanding of this particular problem.

Before leaving the issue of problems in analytical mechanics and Hamiltonian structure, there are two additional comments that merit discussion. A hallmark of the dynamical problem that we have discussed is that the existence of a conservation law fundamentally reduces the number of dimensions needed to describe the spatial and momentum coordinates for the problem. During the latter part of the 18th century, it was widely believed that in three spatial dimensions only a certain class of potentials—identified as *Staeckel potentials* (Goldstein et al., 2002)—would allow there to exist three conservation laws. This gave rise to many fundamental ideas in statistical mechanics during the early years of its development, including the *ergodic hypothesis* that claimed that potentials not meeting the Staeckel conditions would allow orbits to occupy all parts of six-dimensional phase space (three spatial and three momentum coordinates) restricted only by energy (Hamiltonian) conservation. Importantly, Kolmogorov, Arnold, and Moser, or *KAM* (Lichtenberg and Lieberman, 1992; Goldstein et al., 2002), showed in 1954 using perturbation methods related to what we have presented above that the six-dimensional orbits could remain close for a very long time to a Staeckel system that precisely maintained three conservation laws. The insights obtained by KAM allow us to think of a much richer class of problems as having many of the features of an *integrable* problem. This presents a very rich arena of problems that extend far beyond the scope of this text, and the reader is referred to Lichtenberg and Lieberman (1992), especially, and Strogatz (1994) and other treatments of this problem area.

2.3 *Special Functions, Laplacians, and Separation of Variables*

In the solution of the partial differential equations that characterize much of geophysics, as well as related disciplines, we routinely discover that the Laplacian ∇^2 is ubiquitous. In order to address real-world problems arising, it is important that we be able to separately consider the role of the different coordinates upon the solution to the problem. This is especially true in cylindrical and spherical coordinates. The ability to do this and to construct complete basis sets that span the different angle-based geometries makes it possible to solve a wealth of problems. The method of *separation of variables* is universally employed because it exploits an intrinsic feature of these geometries: The independent forms of angular variation are independent of length scale, and the same morphological features apply to environments ranging from the distribution of gravitational and magnetic fields in galactic, stellar, and planetary environments down to atomic and nuclear structures. In the process of obtaining solutions that separate the underlying variables, we encounter a class of second-order ordinary differential equations that belong broadly to the categories that we identified earlier. Another quintessential aspect of geometry emergent from Laplacians in physics is the appearance of $1/r$ potentials, in spherical geometry, whose Laplacian vanishes, and to the emergence of plane waves in different environments, including those in circular and cylindrical geometries. These two considerations allow us to construct a class of expansions utilizing so-called *generating functions*, particularly for Bessel functions, Legendre polynomials, and related quantities, that provide another route to identifying the different coordinate dependencies that prevail in those problems.

We will begin by reviewing for the Laplacian the method of separation of variables in Cartesian, polar, and cylindrical geometries, and spherical coordinates. In each case, we will derive the associated second-order differential equations and explore their solution. We will integrate into that discussion the relevant generating functions that apply, and present a brief overview of some of the methods that can be used. In particular, we will introduce the Bessel function in the context of plane-wave propagation in polar and cylindrical coordinates, and spherical harmonics including both the regular and associated Legendre polynomials in applications utilizing spherical coordinates.

The methodology of separation of variables is based upon the idea of separating the behavior in a given multidimensional problem into the functional dependence of each individual coordinate in that geometry. Once that objective is met, we can determine if the solution types that emerge for each coordinate can provide a complete basis set in that geometry. As a simple paradigm, we will focus on Helmholtz's equation for a monochromatic wave described by the function $\Psi(\boldsymbol{x})$, namely,

$$\nabla^2 \Psi(\boldsymbol{x}) + k^2 \Psi(\boldsymbol{x}) = 0, \tag{2.95}$$

where $\boldsymbol{x}$ is the position in whatever geometry that we chose. Normally, we would expect the solution to have the form of a plane wave, or a linear combination of them. Hence,

$$\Psi(\boldsymbol{x}) = \Psi(\boldsymbol{0}) \exp(\mathrm{i}\boldsymbol{k} \cdot \boldsymbol{x}), \tag{2.96}$$

where $\boldsymbol{k}$, satisfying

$$\boldsymbol{k} \cdot \boldsymbol{k} = k^2, \tag{2.97}$$

is the wavenumber. An even simpler situation, that we will explore in the context of spherical geometries, applies if the Laplacian of Ψ vanishes—*Poisson's equation*. Let us begin with Cartesian geometry.

2.3.1 Cartesian Coordinates and Separation of Variables

Suppose we have a function $\Psi(x, y, z)$ whose Laplacian has some special property. For example, let us suppose that it satisfies Helmholtz's equation:

$$\begin{aligned}
0 &= \nabla^2 \Psi(x, y, z) + k^2 \Psi(x, y, z) \\
&= \frac{\partial^2 \Psi(x, y, z)}{\partial x^2} + \frac{\partial^2 \Psi(x, y, z)}{\partial y^2} + \frac{\partial^2 \Psi(x, y, z)}{\partial z^2} + k^2 \Psi(x, y, z),
\end{aligned} \tag{2.98}$$

and satisfies a specified set of boundary conditions. Let us now presume that the solution can be expressed in the "separable" form

$$\Psi(x, y, z) = X(x)Y(y)Z(z). \tag{2.99}$$

We will now divide this expression by $\Psi(x, y, z)$ and obtain

$$0 = \frac{1}{X(x)} \frac{d^2 X(x)}{dx^2} + \frac{1}{Y(y)} \frac{d^2 Y(y)}{dy^2} + \frac{1}{Z(z)} \frac{d^2 Z(z)}{dz^2} + k^2, \quad (2.100)$$

where we have also replaced partial by ordinary derivatives We now observe a remarkable feature of this result: There are four distinguishable terms, with each of the first three depending solely on one Cartesian coordinate—x, y, or z—and with the last term k^2 being coordinate independent. Accordingly, each of those terms must be independently preserved, that is, constants having no dependence on *any* coordinate since each coordinate is independent. We will refer to the first three constants as $-k_x^2$, $-k_y^2$, $-k_z^2$ and thereby note that

$$0 = k^2 - k_x^2 - k_y^2 - k_z^2. \quad (2.101)$$

We can think of these three constants as defining the relevant eigenmodes for this problem. What is involved now is taking all possible solutions for k_x, k_y, and k_z, as observed before, and constructing appropriate linear combinations of solutions to match the boundary conditions that apply, such as the behavior of Ψ on the boundaries of a box. How will this behavior manifest in other coordinate geometries?

2.3.2 Polar and Cylindrical Coordinates and Separation of Variables; Bessel and Generating Functions

We now consider Helmholtz's equation in two dimensions, that is, polar coordinates, applied to a function $f(r, \theta)$, namely,

$$0 = \nabla^2 f(r, \theta) + k^2 f(r, \theta)$$
$$= \frac{1}{r} \frac{\partial}{\partial r} \left[r \frac{\partial f(r, \theta)}{\partial r} \right] + \frac{1}{r^2} \frac{\partial^2 f(r, \theta)}{\partial \theta^2} + k^2 f(r, \theta), \quad (2.102)$$

which describes a sinusoidal wave propagating away from the origin. (We will not incorporate behavior in the z-direction at this time, but will introduce that aspect when we consider cylindrical coordinates.) Naturally, we expect this wave to decay as it goes to infinity. Since this problem is linear, we can consider a linear combination of all possible *modes*, once we identify how to find them.

As before, we consider a separable solution of the form

$$f(r, \theta) = R(r)\Theta(\theta), \quad (2.103)$$

which we now introduce into (2.102) and obtain

$$0 = \frac{1}{rR(r)} \frac{\mathrm{d}}{\mathrm{d}r} \left[r \frac{\mathrm{d}R(r)}{\mathrm{d}r} \right] + \frac{1}{r^2 \Theta(\theta)} \frac{\mathrm{d}^2 \Theta(\theta)}{\mathrm{d}\theta^2} + k^2. \qquad (2.104)$$

We observe that this form is not completely separated, so we multiply by r^2 and collect like terms to obtain

$$0 = \left\{ \frac{r}{R(r)} \frac{\mathrm{d}}{\mathrm{d}r} \left[r \frac{\mathrm{d}R(r)}{\mathrm{d}r} \right] + r^2 k^2 \right\} + \frac{1}{\Theta(\theta)} \frac{\mathrm{d}^2 \Theta(\theta)}{\mathrm{d}\theta^2}. \qquad (2.105)$$

It follows that both the term in braces, which depends solely upon r, and the concluding term, which depends solely upon θ, must be independent constants. However, the $\Theta(\theta)$ dependence must be periodic in θ and, hence, must satisfy for integer n

$$\frac{1}{\Theta(\theta)} \frac{\mathrm{d}^2 \Theta(\theta)}{\mathrm{d}\theta^2} = -n^2, \qquad (2.106)$$

from which we conclude that

$$\Theta(\theta) = \exp(in\theta)\Theta(0) \qquad (2.107)$$

and

$$0 = \frac{r}{R(r)} \frac{\mathrm{d}}{\mathrm{d}r} \left[r \frac{\mathrm{d}R(r)}{\mathrm{d}r} \right] + r^2 k^2 - n^2. \qquad (2.108)$$

We identify this nth radial eigenfunction as $R_n(r)$ and observe

$$0 = r^2 \frac{\mathrm{d}^2 R_n(r)}{\mathrm{d}r^2} + r \frac{\mathrm{d}R_n(r)}{\mathrm{d}r} + [r^2 k^2 - n^2]R_n(r). \qquad (2.109)$$

Finally, we render the equation dimensionless by replacing kr with r and obtain *Bessel's equation*

$$0 = r^2 \frac{\mathrm{d}^2 J_n(r)}{\mathrm{d}r^2} + r \frac{\mathrm{d}J_n(r)}{\mathrm{d}r} + [r^2 - n^2]J_n(r), \qquad (2.110)$$

for *Bessel functions* of the first kind and order n.

Power series solutions are generally obtained by expressing the solution as an infinite series with undetermined coefficients and employing the differential equation to establish a relationship between the coefficients. When combined with the initial conditions, that is, the function value (or its nth derivative) and its first (or $n+1$st) derivative at $r = 0$, the value of all coefficients can be uniquely determined. This topic is developed in detail in elementary treatments such as Boas (2006), Butkov (1968), and Greenberg (1998), as well as more advanced ones such as Mathews and Walker (1970), and is often referred to as the *method of Frobenius.* It is important to note that such series expansions do not always converge, and it is important for users of these methods to check for that.

Series solutions can now be expressed

$$J_n(r) = \frac{r^n}{2^n n!} \left\{ 1 - \frac{r^2}{2(2n+2)} + \frac{r^4}{2 \cdot 4(2n+2)(2n+4)} - \cdots \right\},$$

(2.111)

which is especially useful for small values of r. Note that this class of solution is chosen to be regular at the origin.

Bessel functions of the second kind, denoted $Y_n(r)$, diverge logarithmically at the origin, and correspond to solutions of Bessel's equation that are irregular at the origin. Note, however, that their singular behavior at the origin is an outcome of the coordinate singularity at the origin. They appear, for example, in problems with annular geometry. We plot in Figure 2.5 some low-order Bessel functions of the first kind.

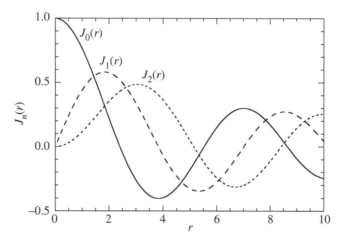

Figure 2.5.   Low-order Bessel functions.

A couple of examples include

$$J_0(r) = 1 - \frac{r^2}{2^2} + \frac{r^4}{2^2 4^2} - \cdots$$

$$J_1(r) = \frac{r}{2} - \frac{r^3}{2^2 \cdot 4} + \frac{r^5}{2^2 \cdot 4^2 \cdot 6} - \cdots.$$

(2.112)

Returning to our expected solution having the form $\exp(i\mathbf{k} \cdot \mathbf{x})$, it is useful to explore the nature of the quantity $\exp[ir\cos(\theta)]$, where θ is the angle between $\mathbf{k}$ and $\mathbf{x}$. We now construct a *generating function* for the Bessel function using the forms

$$\exp\left[\tfrac{1}{2}r(t - 1/t)\right] = \sum_{n=-\infty}^{\infty} t^n J_n(r)$$

$$\exp[ir\cos(\theta)] = \sum_{n=-\infty}^{\infty} i^n \exp(in\theta)J_n(r), \qquad (2.113)$$

where we employed $t = i\exp(i\theta)$ to derive the second formulation from the first. The latter expression provides the rationale for our designating this as a "generating" function: The series that emerges decouples the different azimuthal modes, that is, modes that repeat n times over 2π angular evolution. As exercises, we will show how the generating function can be employed—by taking derivatives with respect to r and t—to derive the series and so-called recurrence formulae for the Bessel function. We will not consider Bessel functions of order equal to half an odd integer—these are related to spherical Bessel functions and are important to normal modes of the Earth—and modified Bessel functions that emerge when k^2 is negative, which are treated in depth in many references (Watson, 1995; Mathews and Walker, 1970; Whittaker and Watson, 1979; Jeffreys and Jeffreys, 1999).

Let us now proceed to explore how these principles impact three dimensions. In particular, if we add a $\partial^2/\partial z^2$ term to our Laplacian in polar coordinates, our separation of variables method (2.105) appends an additional term in

$$\frac{1}{Z(z)} \frac{d^2 Z(z)}{dz^2},$$

which in turn adds a constant to the $-n^2$ emerging from the θ variable. In those situations, especially if it produces an exponential in contrast with sinusoidal variation in the z-direction can give rise to yet another class of Bessel function, known as the *Hankel function*, which is a linear combination of the Bessel functions J_n and Y_n. The references cited above should be consulted regarding the behavior of these related quantities, which are relatively uncommon in geophysical applications.

Finally, there are circumstances where we need to evaluate an approximate value for $J_n(r)$ for large r. Returning to (2.110), we observe that the first derivative term obfuscates the asymptotic properties of the solution. We eliminate the first derivative term by making the substitution

$$J_n(r) = \frac{v_n(r)}{\sqrt{r}}, \tag{2.114}$$

and obtain that

$$\frac{d^2 v_n(r)}{dr^2} + \left(1 - \frac{n^2 - \frac{1}{4}}{r^2}\right) v_n(r) = 0, \tag{2.115}$$

which has Sturm–Liouville form. This equation can readily be solved using JWKB methods, for example, especially for large r where a sinusoidal behavior including a phase shift can be introduced: This approach relates to what is called the *method of stationary phase*, which we shall describe later. This is a very rich problem area, and Watson (1995) provides an exhaustive compilation of fundamental results. Mathews and Walker (1970) provide a succinct summary of results, including the asymptotic

$$J_n(r) \approx \sqrt{\frac{2}{\pi r}} \cos\left(r - \frac{n\pi}{2} - \frac{\pi}{4}\right). \tag{2.116}$$

In practical applications, there are situations where it becomes necessary to link the series and asymptotic expansions together, the so-called *connection problem*. A useful method for doing this is to establish an approximation for the error term in each of the two expansions, and then identify whether either or both are sufficiently small for the purpose at hand and transition from one expansion to the other accordingly. The NIST handbook (Olver, 2010) can be helpful in this regard. A special case of (2.115) occurs when $n = \pm 1/2$, where the solution is trivial to obtain and is often called a *spherical Bessel function*. This is defined in Mathews and Walker (1970) and Olver (2010).

Finally, it is worth noting that k is often an eigenvalue for the problem associated with a particular set of boundary conditions, for example, in exploring the different azimuthal (i.e., angular) modes of a vibrating drum head, which requires that the amplitude of the vibration vanish at the edge of the drum. In order for this to happen, k must be selected to assure that $J_n(kR)$, where R is the radius of the drum, vanishes thereby making the roots of the Bessel function of particular importance. The transition to cylindrical coordinates is relatively simple since the z-dependence and the $\partial^2/\partial z^2$ terms, as in the Cartesian problem, are immediately separable. The eigenvalue associated with the z-direction must again be selected to conform with the boundary conditions that emerge. This in turn affects the radial eigenvalue that we have already discussed. There are other circumstances, for example, relating to torsional waves in the Earth's core, where there is no explicit boundary condition present, but it is necessary and sufficient for the solution to be bounded.

Our focus in this section was on the emergence of Bessel functions and their relatives in a broad class of problems bearing polar or cylindrical symmetry. However, we should note that the $\exp(in\theta)$ terms are azimuthal or meridional eigenfunctions and that this geometrical character is universal. For any well-behaved function $f(\theta)$, we can express it using a Fourier-like decomposition

$$f(r,\theta) = \sum_{n=-\infty}^{\infty} f_n(r) \exp(in\theta), \qquad (2.117)$$

where

$$f_n(r) = \frac{1}{2\pi} \int_0^{2\pi} f(r,\theta) \exp(-in\theta)\, d\theta. \qquad (2.118)$$

The ability to exploit the existence of eigenfunctions for Laplace's equation in polar and cylindrical geometry can be extended to three dimensions in spherical geometry owing to the completeness of the basis sets produced. This will be an essential aspect of our next section.

2.3.3 Spherical Coordinates and Separation of Variables; Green's and Generating Function; Spherical Harmonics

As a motivation for determining the solution to Poisson's, Helmholtz's, and Laplace's equations in three-dimensional geometry, let us consider the relation between the potential of a planet, such as the Earth, and its internal density distribution. (A formally identical problem emerges in magnetospheric physics attendant to determining the Earth's magnetic field.) We will exploit the properties of spherical geometry via the method of separation of variables and the use of spherical harmonics. We will then employ that as a springboard for studying the issue of Green's functions for gravitational (and magnetostatic) problems in chapter 4, which addresses partial differential equations.

2.3.3.1 Helmholtz's and Laplace's Equation

From Newton's universal law of gravitation, we learned that the potential energy function can be expressed

$$\Phi(\boldsymbol{x}) = -G \int_V \frac{\rho(\boldsymbol{x}')}{|\boldsymbol{x}' - \boldsymbol{x}|}\, d^3x', \qquad (2.119)$$

where G is the universal gravitational constant, approximately $6.67384 \pm 0.00080 \times 10^{-11}\,\text{m}^3\,\text{kg}^{-1}\,\text{s}^{-2}$, and V is the volume over which the mass is distributed, and where we have omitted making explicit reference to time. This expression describes how the potential energy at $\boldsymbol{x}$ depends on contributions of mass at all possible positions $\boldsymbol{x}'$. A common geophysical problem emerges from the observation that $\Phi(\boldsymbol{x})$ is known, from the orbits of artificial satellites, and we wish to establish the density distribution $\rho(\boldsymbol{x})$ that produced it. This expression, in turn, is a form of *integral equation*. To understand this, we need to appreciate the properties of $|\boldsymbol{x}' - \boldsymbol{x}|^{-1}$.

Since this latter quantity enjoys spherical symmetry, consider now the properties of the radial distance r in the context of differential operators. We recall, for spherical geometry, that the Laplacian (1.44) of any scalar function f has the form

$$\nabla^2 f \to \frac{1}{r^2}\frac{\partial}{\partial r}\left(r^2 \frac{\partial f}{\partial r}\right). \qquad (2.120)$$

It follows immediately, for $f = 1/r$, that

$$\nabla^2\left(\frac{1}{r}\right) = \frac{1}{r^2}\frac{\partial}{\partial r}(-1) = 0 \qquad (2.121)$$

so long as $r \neq 0$. This is the simplest solution to *Poisson's equation*. However, when $r = 0$, the latter expression is not well defined. Therefore, it is important to evaluate

$$\int_V \nabla^2\left(\frac{1}{r}\right) d^3r = \int_S \nabla\left(\frac{1}{r}\right)\cdot \hat{\boldsymbol{n}}\, d^2r = -\int_S \left(\frac{1}{r^2}\right) d^2r, \quad (2.122)$$

where V is any volume that contains the origin and S is the containing surface. Since we have already established that there is no contribution to this integral emerging from points away from the origin, let us consider a volume V' that corresponds to a sphere with a vanishingly small radius r, and that d^2r reduces to $r^2\, d\Omega$, where $d\Omega$ is the solid angle $\sin\theta\, d\theta\, d\varphi$. Pulling this together, we then obtain

$$\int_{V'} \nabla^2\left(\frac{1}{r}\right) d^3r = -4\pi. \qquad (2.123)$$

Evidently, $\nabla^2 r^{-1}$ must be both negative and infinite as $r \to 0$. We will now explore the action of the Laplacian on the potential $\Phi(x)$ in Eq. (2.119).

We begin, then, by taking the Laplacian, namely,

$$\nabla^2\Phi(x) = -G\int_V \rho(x')\nabla^2\left(\frac{1}{|x'-x|}\right)d^3x'$$
$$\to -G\int_{V'} \rho(x')\nabla^2\left(\frac{1}{|x'-x|}\right)d^3x', \qquad (2.124)$$

where V' describes an infinitesimally small sphere for x' around x. Accordingly, we can extract $\rho(x)$ from inside the integral and obtain

$$\nabla^2\Phi(x) \to -G\rho(x)\int_{V'} \nabla^2\left(\frac{1}{|x'-x|}\right)d^3x' = 4\pi G\rho(x),$$
$$(2.125)$$

which is sometimes referred to as *Poisson's equation*. Given our derivation, it is evident that the solution to Poisson's equation is given by the potential equation (2.119). Importantly, we observe that the effect of taking the Laplacian inside the integral was to identify and extract the remaining portion of the integrand multiplied by a minus sign and 4π, emerging from the solid angle integration. Accordingly, we write

$$\nabla^2\left(\frac{1}{r}\right) = -4\pi\delta(r), \qquad (2.126)$$

where we refer to $\delta(x)$ as the *Dirac δ function* which has two fundamental features:

$$\delta(x) = 0 \quad \text{if } |x| = 0, \qquad (2.127)$$

and

$$f(y) = \int_V f(x)\delta(x-y)\,d^3x, \qquad (2.128)$$

revealing that the δ function is singular and behaves inside integrals in a special way. It offers no contribution away from $\mathbf{0}$ but picks up exactly one unit of the remaining part of the integrand. Recalling our definition of the Green's function (2.40), it follows that the Green's function for the Laplacian is

$$G(x,x') = -\frac{1}{4\pi|x-x'|}. \qquad (2.129)$$

Before proceeding to explore the Laplacian in spherical geometry, it is important to remember a fundamental feature of potential distributions, for example, for gravitation or magnetostatics. We note that $1/r$ for $r > 0$ is a solution to Laplace's equation,

$$\nabla^2 \frac{1}{r} = \frac{1}{r^2} \frac{\partial}{\partial r} \left[r^2 \frac{\partial (1/r)}{\partial r} \right] = 0. \tag{2.130}$$

Moreover, we also note that we can similarly obtain a spherically symmetric solution to Helmholtz's equation, that is, $\exp(ikr)/r$, since

$$\nabla^2 \left[\frac{\exp(ikr)}{r} \right] + k^2 \left[\frac{\exp(ikr)}{r} \right]$$
$$= \frac{1}{r^2} \frac{\partial}{\partial r} \left\{ r^2 \frac{\partial [\exp(ikr)/r]}{\partial r} \right\} + k^2 \left[\frac{\exp(ikr)}{r} \right]$$
$$= 0, \tag{2.131}$$

after some algebra. Accordingly, it is also easy to show that the Green's function for Helmholtz's equation is given by

$$G(\boldsymbol{x}, \boldsymbol{x}') = -\frac{\exp(ik|\boldsymbol{x} - \boldsymbol{x}'|)}{4\pi |\boldsymbol{x} - \boldsymbol{x}'|}. \tag{2.132}$$

It is noteworthy that the Bessel function solution (2.114) to Helmholtz's equation in two dimensions (2.116) is sinusoidal and falls off as $r^{-1/2}$ while that in three dimensions as noted here is also sinusoidal but falls off as r^{-1}.

2.3.3.2 *From Green's Function to Generating Function*

Suppose you have a distribution of mass where you know each point's location $\boldsymbol{r}'$ measured relative to the center of that mass distribution. Moreover, you are an observer situated at $\boldsymbol{r}$, which is further away from the coordinate origin than any point $\boldsymbol{r}'$ in that distribution. The potential energy associated with that point will then vary as

$$\frac{1}{|\boldsymbol{r} - \boldsymbol{r}'|} = \frac{1}{\sqrt{r^2 - 2\boldsymbol{r} \cdot \boldsymbol{r}' + r'^2}} = \frac{1}{r\sqrt{1 - 2t\mu + t^2}}, \tag{2.133}$$

where $t = r'/r < 1$ and $\mu = \cos(\theta)$, where θ is the angle between $\boldsymbol{r}$ and $\boldsymbol{r}'$. Note that the symbol μ is commonly used for $\cos(\theta)$ above. It is useful to decouple the angular dependence present from the various powers of t, that is, of $r'/r < 0$, which we will designate as providing the different *multipole* components of the expansion. We call such a decoupling of terms a *generating function* since it "generates" the angular dependence present for each multipole term.

We now define (2.133) as the generating function for *Legendre polynomials* according to

$$\frac{1}{r\sqrt{1 - 2t\mu + t^2}} = \frac{1}{r}\sum_{\ell=0}^{\infty} t^\ell P_\ell(\mu), \tag{2.134}$$

where $P_\ell(\mu)$ is the Legendre polynomial of order ℓ. Commonly, in treatments of the Legendre polynomial, the symbol x is employed in place of μ. The first few examples are shown in the Figure 2.6.

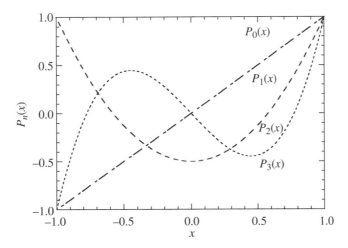

Figure 2.6.   Low-order Legendre polynomials.

It is easy to show that

$$0 = (1 - x^2)\frac{d^2 P_\ell(x)}{dx^2} - 2x\frac{dP_\ell(x)}{dx} + \ell(\ell + 1)P_\ell(x)$$

$$= \frac{d}{dx}\left[(1 - x^2)\frac{dP_\ell(x)}{dx}\right] + \ell(\ell + 1)P_\ell(x). \tag{2.135}$$

It is possible to show by induction that the Legendre polynomials are explicitly given by *Rodrigues's formula* (Mathews and Walker, 1970; Jackson, 1999) for $\ell \geqslant 0$:

$$P_\ell(x) = \frac{(-1)^\ell}{2^\ell \ell!}\frac{d^\ell}{dx^\ell}(1 - x^2)^\ell, \tag{2.136}$$

with some examples being

$$P_0(x) = 1$$
$$P_1(x) = x$$
$$P_2(x) = \frac{1}{2}(3x^2 - 1)$$
$$P_3(x) = \frac{x}{2}(5x^2 - 3). \tag{2.137}$$

As in the case of the generating function for the Bessel function, the generating function for Legendre polynomials can be employed to generate recurrence formulae connecting different orders as well as derivatives of the different polynomials. Furthermore, we can employ these relationships to derive the *orthogonality* of the Legendre polynomials, namely,

$$\int_{-1}^{1} P_\ell(x) P_m(x)\, dx = \begin{cases} 0 & \text{if } \ell \neq m, \\ \dfrac{2}{2\ell + 1} & \text{if } \ell = m. \end{cases} \tag{2.138}$$

We will utilize some of these relationships in the exercises at the end of this chapter. Now, we wish to explore the use of separation of variables in application to the Laplacian in spherical geometry to complete this analysis.

2.3.3.3 *Separation of Variables in Spherical Geometry*

Let us now consider the Laplacian of $\Psi(r, \theta, \phi)$ as described in chapter 1, where we now separate the variables according to

$$\Psi(r, \theta, \phi) = R(r)\Theta(\theta)\Phi(\phi), \tag{2.139}$$

so that

$$\nabla^2 \Psi(r, \theta, \phi) = \frac{1}{r^2} \frac{\partial}{\partial r} \left[r^2 \frac{\partial \Psi(r, \theta, \phi)}{\partial r} \right]$$
$$+ \frac{1}{r^2 \sin \theta} \frac{\partial}{\partial \theta} \left[\sin \theta \frac{\partial \Psi(r, \theta, \phi)}{\partial \theta} \right]$$
$$+ \frac{1}{r^2 \sin^2 \theta} \frac{\partial^2 \Psi(r, \theta, \phi)}{\partial \phi^2}. \tag{2.140}$$

In these applications, θ describes the angle between r and the z-axis, rendering $z = r \cos(\theta) = r\mu$, with x and y corresponding to $r \sin(\theta) \cos(\phi)$ and $r \sin(\theta) \sin(\phi)$, respectively. We divide both sides by $\Psi(r, \theta, \phi)$ for the homogeneous case and adopt separable form to obtain

$$
\begin{aligned}
0 &= \frac{\nabla^2 \Psi(r, \theta, \phi)}{\Psi(r, \theta, \phi)} \\
&= \frac{1}{r^2 R(r)} \frac{d}{dr}\left[r^2 \frac{dR(r)}{dr} \right] + \frac{1}{r^2 \sin \theta \Theta(\theta)} \frac{d}{d\theta}\left[\sin \theta \frac{d\Theta(\theta)}{d\theta} \right] \\
&\quad + \frac{1}{r^2 \sin^2 \theta \Phi(\phi)} \frac{d^2 \Phi(\phi)}{d\phi^2}.
\end{aligned} \tag{2.141}
$$

We now multiply this quantity by $r^2 \sin^2 \theta$ and factor terms accordingly:

$$
\begin{aligned}
0 = \sin^2 \theta &\left[\left\{ \frac{1}{R(r)} \frac{d}{dr}\left[r^2 \frac{dR(r)}{dr} \right] \right\} \right. \\
&\quad \left. + \left\{ \frac{1}{\sin \theta \Theta(\theta)} \frac{d}{d\theta}\left[\sin \theta \frac{d\Theta(\theta)}{d\theta} \right] \right\} \right] \\
&\quad + \frac{1}{\Phi(\phi)} \frac{d^2 \Phi(\phi)}{d\phi^2}.
\end{aligned} \tag{2.142}
$$

We observe that the only ϕ-dependence arises from the last term. We further observe that $\Phi(\phi)$ must be periodic over 2π, as we noted previously in polar and cylindrical geometry. It follows that the term in the Laplacian in ϕ must be $-n^2$ for $n = 0, \pm 1, \pm 2, \ldots$. This corresponds to

$$
\Phi(\phi) = \exp(in\phi). \tag{2.143}
$$

We express this now as

$$
\begin{aligned}
0 = &\left\{ \frac{1}{R(r)} \frac{d}{dr}\left[r^2 \frac{dR(r)}{dr} \right] \right\} \\
&+ \left\{ \frac{1}{\sin \theta \Theta(\theta)} \frac{d}{d\theta}\left[\sin \theta \frac{d\Theta(\theta)}{d\theta} \right] - \frac{n^2}{\sin^2 \theta} \right\}. \tag{2.144}
\end{aligned}
$$

We observe that the first set of braces contains the only terms in r while the second set contains the only terms in θ, and, therefore, the two quantities in braces must be independent constants. To simplify the algebra, we now substitute x for $\cos \theta$—as is typically done in most applied mathematics texts—and regard

Θ as a function of x, whereupon we can write

$$0 = \left\{ \frac{1}{R(r)} \frac{d}{dr} \left[r^2 \frac{dR(r)}{dr} \right] \right\}$$
$$+ \left\{ \frac{1}{\Theta(x)} \frac{d}{dx} \left[(1-x^2) \frac{d\Theta(x)}{dx} \right] - \frac{n^2}{1-x^2} \right\}. \quad (2.145)$$

We observe that the two sets of braces have segregated the r and the x (or $\cos\theta$) dependence. Since both terms in braces must now be equal and opposite constants, we must investigate the conditions under which a solution to the two eigenmode problems exists. Owing to the homogeneity of the radial term $R(r)$, we expect a power-law solution of the form r^ℓ and observe the associated constant to be $\ell(\ell+1)$; however, at this stage, there is no restriction on the value of ℓ. We must now identify for what values of ℓ a solution for the Θ eigenmode problem exists for $-1 \leqslant x \leqslant 1$. Accordingly, we now write

$$\frac{d}{dx} \left[(1-x^2) \frac{d\Theta(x)}{dx} \right] - \frac{n^2}{1-x^2} \Theta(x) + \ell(\ell+1)\Theta(x) = 0. \quad (2.146)$$

For the case $n = 0$, we observe that this is the same differential equation that we encountered in (2.135) for $P_\ell(x)$, assuming that ℓ is a positive integer. Importantly, if we replace the positive-valued integer ℓ by $-\ell - 1$, we also recover the same differential equation. Is integer ℓ a sufficient condition to guarantee that a regular solution for $\Theta(x)$ exists? It is easy to show for integer $\ell \geqslant 0$ that a polynomial solution of order ℓ exists; however, if ℓ is not an integer, the series does not terminate and yields a solution that diverges near the origin. We will not provide a proof here—one can be found in Jackson (1999)—but simply state the solution to (2.135), namely, the associated Legendre functions $P_\ell^n(x)$, by exploiting Rodrigues's formula (2.136) and defined by

$$P_\ell^n(x) = \frac{(-1)^n}{2^\ell \ell!} (1-x^2)^{n/2} \frac{d^{\ell+n}}{dx^{\ell+n}} (x^2 - 1)^\ell, \quad (2.147)$$

valid for $\ell \geqslant 0$ and $|n| \leqslant \ell$.

Recalling that $x = \cos\theta$, it follows that $(1-x^2)^{1/2}$ is equivalent to $\sin\theta$ and we give here some examples of the associated Legendre polynomials, namely,

$$P_1^1(x) = (1-x^2)^{1/2}$$
$$P_2^1(x) = 3x(1-x^2)^{1/2}$$
$$P_2^2(x) = 3(1-x^2). \quad (2.148)$$

There is, as before, an orthogonality condition for the associated Legendre functions, namely,

$$\int_{-1}^{1} P_{\ell}^{n}(x)P_{m}^{n}(x)\,\mathrm{d}x = \begin{cases} 0 & \text{if } \ell \neq m, \\ \dfrac{2}{2\ell + 1}\dfrac{(\ell + n)!}{(\ell - n)!} & \text{if } \ell = m. \end{cases} \tag{2.149}$$

Note that we have not expressed the orthogonality condition that applies if the integrand contains associated Legendre functions with different azimuthal orders, that is, different values of n. In that instance, the azimuthal eigenmodes $\exp(in\phi)$ will be orthogonal.

In order to complete our description of solutions to the homogeneous Laplace equation in spherical geometry, we must now combine the eigenmodes for the θ and ϕ behavior. The literature, unfortunately, has several competing and not always consistent notations. We will employ the spherical harmonics $Y_{\ell n}(\theta, \phi)$ defined in Jackson (1999), which are widely employed by the physics community, including space plasma physicists. There are two notations employed commonly by solid-earth physicists, and these are described fully in Stacey and Davis (2008). Kaula (1968) provides a comprehensive introduction to the use of spherical harmonics in geophysics, and pioneered their application to satellite geodesy. Accordingly, we employ the fully normalized functions

$$Y_{\ell n}(\theta, \phi) = \sqrt{\frac{2\ell + 1}{4\pi}\frac{(\ell - n)!}{(\ell + n)!}}P_{\ell}^{n}(\cos\theta)\exp(in\phi). \tag{2.150}$$

Finally, we wish to point out that the spherical harmonics $Y_{\ell n}(\theta, \phi)$ in three dimensions, just like azimuthal harmonics $\exp(in\theta)$ in two, constitute a complete basis set. In other words, we can express essentially any function $f(r, \theta, \phi)$ in the form

$$f(r, \theta, \phi) = \sum_{\ell=0}^{\infty}\sum_{n=-\ell}^{\ell} f_{\ell n}(r)Y_{\ell n}(\theta, \phi), \tag{2.151}$$

where the radial modes satisfy

$$f_{\ell n}(r) = \int_{\theta=0}^{\pi}\int_{\phi=0}^{2\pi} f(r, \theta, \phi)Y_{\ell n}^{*}(\theta, \phi)\sin\theta\,\mathrm{d}\theta\,\mathrm{d}\phi, \tag{2.152}$$

by virtue of the orthogonality condition

$$\int_{\theta=0}^{\pi}\int_{\phi=0}^{2\pi} Y_{\ell n}(\theta, \phi)Y_{mk}^{*}(\theta, \phi)\sin\theta\,\mathrm{d}\theta\,\mathrm{d}\phi = \delta_{\ell m}\delta_{nk}. \tag{2.153}$$

We will later observe that the completeness of the spherical harmonics allows us to construct Green's functions for many problems in spherical coordinates. Spherical harmonic structure is preserved under coordinate rotation and is intricately linked to the *rotation group*. Importantly, spherical harmonic expansions in one coordinate system can be converted into equivalent spherical harmonic expansions in a rotated coordinate system; this illustrates the existence of *addition theorems* for spherical harmonics in general as well as Legendre polynomials—see, for example, Whittaker and Watson (1979) as well as Olver (2010). This topic resides in the province of partial differential equations and will be addressed in chapter 4.

Before departing from our discussion of the Legendre orthogonal polynomials, it is worth noting that there are a number of other types of orthogonal polynomials that find other applications. Arfken and Weber (2005) present a listing of several of these, including the Hermite, Laguerre, and Chebyshev polynomials, in addition to the Legendre. Each of these satisfies a form of Rodrigues's formula and has generating functions. Moreover, as described in Olver (2010), they also possess orthogonality properties under a specific weighting factor. In the case of Legendre polynomials, the weighting factor is simply 1 and approximating a function using Legendre polynomial expansions minimizes the integrated square-error over the interval $[-1, 1]$. Of these, the Chebyshev polynomials are often employed in geophysical fluid dynamics simulations (Glatzmaier, 2013). They are simpler to use than Legendre polynomials and in many circumstances minimize the maximum error observed in approximating a function over some interval (Cheney, 1982). Other polynomials of this type appear in quantum mechanical applications for hydrogenic atoms—*Laguerre polynomials*—and quantum mechanical harmonic oscillators—*Hermite polynomials*. These issues are beyond the scope of this text.

We will return later to questions pertaining to perturbation methods as we turn our focus to couple nonlinear ordinary differential equations and the special role that they occupy in geophysics. Nonlinear first-order problems abound in geophysics. Historically, the impetus for treating these mathematical questions had their origin in problems ranging from the motion of the Moon to the nature of friction. There are relatively few exact solutions available, so we will focus on one prominent example that is relevant to geophysical fluid dynamics, and a coupled equation problem that has been especially important in geophysics and the birth of chaos theory.

2.4 *Nonlinear Ordinary Differential Equations*

We saw earlier how a coupled pair of first-order equations could be employed to describe Newton's laws of motion, and that the results included a conservation law as well as a complete solution via a quadrature. Could this kind of behavior be generic, or could they support behavior that was chaotic? In the late 1960s, this was seen as an important question in trying to understand the nature of the Earth's dynamo, which manifests quasiperiodic behavior, magnetic field reversals, and other complex behaviors. It was thought that a better understanding of the Earth could be obtained by developing a pair of coupled equations to describe the interaction of the Earth's rotational motion, which we shall denote as x, and the electric current produced by its field, which we shall denote by y. The so-called Bullard homopolar dynamo had the functional form, where f and g are scalar functions,

$$\dot{x} = f(x, y)$$
$$\dot{y} = g(x, y). \qquad (2.154)$$

Both f and g had quadratic dependence upon x and y, although the latter is not critical to our argument. Importantly, f and g were such that neither x nor y would become unbounded. Nevertheless, however sophisticated the models that were developed, no one succeeded in producing any kind of behavior that looked like that associated with the Earth.

2.4.1 Bullard's Homopolar Dynamo

Bullard (1955) described a homopolar dynamo as "one in which a conductor moves steadily in a constant magnetic field and produces a direct current without the use of a commutator."

Consider a disk that rotates about its axis in a constant field parallel to the axis as shown in Figure 2.7. Current is produced by two sliding contacts (called "brushes" in the original model), one of which rubs the outside of the disk and the other on the axle about which the disk rotates. The current from the brushes

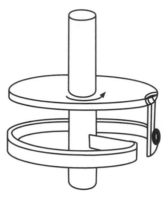

Figure 2.7. Bullard's homopolar dynamo.

is passed through a coil as shown and produces the magnetic field, rendering the dynamo "self-exciting." Bullard developed a circuit model, loosely analogous to an LRC circuit, that incorporates the mutual inductance M of the disc and the coil and the electromotive force induced across the disk $\omega MI/2\pi$, where I is the current through the coil that we introduced earlier as $\dot{q}$ and ω is the angular velocity of the disk. Cook (1973) provides a more detailed derivation for this problem. Accordingly,

$$L\frac{dI}{dt} + RI = \frac{1}{2\pi}\omega MI. \tag{2.155}$$

The mechanical equation describes the balance between the angular acceleration of the disc and the applied torques. Following Cook, we will call the mechanical torque driving the disc G and the resistive mechanical torque caused by the disc-field interactions $MI^2/2\pi$, thereby yielding

$$C\frac{d\omega}{dt} = G - \frac{MI^2}{2\pi}, \tag{2.156}$$

where C is the moment of inertia of the disk. (In this application, I and ω occupy the role of the generic variables x and y given previously.) We observe that there exists a "steady solution" for this pair of equations, that is, one where both time derivatives vanish, when

$$\omega = \frac{2\pi R}{M}$$

$$I = \sqrt{\frac{2\pi G}{M}}. \tag{2.157}$$

Let us identify these steady solutions as ω_0 and I_0, respectively, and consider perturbation expansions of the form

$$\omega = \omega_0 + \delta\omega$$
$$I = I_0 + \delta I. \tag{2.158}$$

We introduce these into (2.155) and (2.156) and keep, after the zero-order terms have cancelled out, only terms of order $\delta\omega$ and δI, and observe that

$$\frac{d\delta\omega}{dt} = -\frac{1}{C}\sqrt{\frac{2MG}{\pi}}\,\delta I$$

$$\frac{d\delta I}{dt} = +\frac{M}{\pi C}\sqrt{\frac{G}{L}}\,\delta\omega. \tag{2.159}$$

We observe, if we were to take the time derivative of these equations, that the resulting second-order equations would have the same form as the simple harmonic oscillator and the resulting phase portrait would be that of a center. What can we say about the nonlinear problem?

When we divide (2.155) by I, we observe that the expressions can be factored and that we can write

$$\frac{1}{IC}\left(G - \frac{MI^2}{2\pi}\right)\frac{dI}{dt} = \frac{1}{L}\left(\frac{\omega M}{2\pi} - R\right)\frac{d\omega}{dt}, \tag{2.160}$$

demonstrating that the equations are integrable and, therefore, we obtain a constant C' from

$$C' = G\ln(I) - \frac{M}{4\pi}I^2 - \frac{M}{4\pi L}\omega^2 + \frac{R}{L}\omega. \tag{2.161}$$

Bullard (1955) explores the time evolution of this system, but the important features of this problem reside in this latter quantity: The system is integrable, which is a much stronger statement than claiming the solution is stable. Moreover, without the necessity of explicitly calculating the time-dependent solution for this problem, as Bullard did, we have demonstrated that it is periodic. What insight can we gain from this that is relevant to general equations of the form (2.154)?

2.4.2 Poincaré–Bendixson Theorem and the Van der Pol Oscillator

Let us consider how this general system of equations must evolve, and let us identify all values of x and y such that the right-hand sides of (2.154) vanish, that is, find all steady or stationary solutions. Suppose our initial conditions are close to a stationary solution, but also suppose that the stationary solution is unstable and "repels" the trajectory as in a saddle point. (If the stationary solution is stable, then the solution will approach it exponentially fast.) So, the solution must evolve away from the stationary solution. However, the solution can never cross itself, because the solution is unique—if it were to cross itself, it would have to repeat itself and would be periodic. (That situation corresponds mathematically to what we obtained for the particle motion problem.) Moreover, the solution cannot run away to infinity, and therefore must spiral toward what appears to be a periodic solution, a so-called *limit cycle*. This, in essence, is a demonstration of the *Poincaré–Bendixson theorem* (Strogatz, 1994; Nayfeh and Balachandran, 1995). This theorem establishes that chaos can *not* occur in a two-dimensional phase plane. The precise statement of the theorem—see Strogatz (1994), p. 203, or Coddington and Levinson (1984), pp. 391–392—presents the formal basis for the observation that chaos can only emerge in systems with three or more governing equations.

All dynamo models that were developed based on two differential equations either were exactly periodic or gradually became periodic in the sense of limit cycles. The resolution to this dilemma was that more than two coupled equations were needed in order to address these questions. Before moving on to consider systems with more than two governing equations, let us consider what has become a paradigm for limit-cycle behavior, the Van der Pol equation.

Let us consider a differential equation associated with vacuum-tube technology, named after its inventor, Balthasar van der Pol, that has a dynamically active resistive component in what is otherwise an LRC circuit. It is commonly written as

$$0 = \ddot{x}(t) + \mu[x^2(t) - 1]\dot{x}(t) + x(t). \qquad (2.162)$$

We observe that the (normalized) resistance R has been replaced by $\mu[x^2(t) - 1]$, a consequence of the appearance of an "active device" (Strogatz, 1994). Let us perform a linearized analysis of this equation, keeping only linear terms, that is,

$$0 = \ddot{x}_0(t) - \mu\dot{x}_0(t) + x_0(t). \qquad (2.163)$$

Looking for linearized solutions of the form $\exp(\nu t)$, we observe that

$$\nu = \frac{\mu \pm \sqrt{\mu^2 - 4}}{2}. \qquad (2.164)$$

We observe three cases:

1. $\mu = 0$ yields purely oscillatory solutions such as those encountered in the simple harmonic oscillator;

2. $\mu < 0$ yields damping, and possibly oscillatory solutions, such as those encountered in the usual LRC circuit; and

3. $\mu > 0$ yields growing, and possibly oscillatory, solutions.

The latter case is of particular interest because the presence of the $x^2(t)$ term in the original equation can modify the sign of the effective resistivity, curbing growth. We associate, as we did in Eq. (2.20), the change in sign of μ with a Hopf bifurcation. We illustrate in Figure 2.8 the behavior for typical initial conditions that is observed for $\mu = \pm 0.5$.

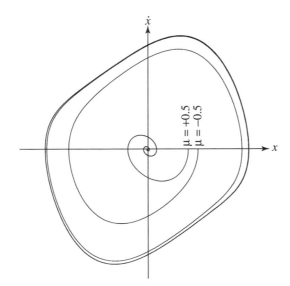

Figure 2.8. Van der Pol oscillator.

Accordingly, we observe damping to the origin, or asymptotic growth culminating in a limit cycle.

The Poincaré–Bendixson theorem allows us to understand what is happening when the stationary solution is not stable. The solution in the unstable case grows until nonlinearity in the underlying equations precludes further growth. At the same time, given that the solution is bounded and is unique—meaning that the phase trajectory can never cross itself—the solution's growth is curbed and in the limit of infinite time assumes the form of a cycle. The Poincaré–Bendixson theorem is remarkable inasmuch as it describes all possible outcomes from coupled pairs of nonlinear ordinary differential equations. It is evident, moreover, that the allowed behaviors are stable, albeit in different ways, and present no sensitivity to the initial conditions. In order to obtain chaotic behavior, more complicated dynamics are essential.

2.4.3 Lorenz Attractor, Perturbation Theory, and Chaos

From a physics-based standpoint, more than two equations are necessary to describe such complex character. Geophysical fluid dynamics seeks to describe complex flows in a variety of environments, including turbulent settings. This feature was recognized by Lorenz, who developed a very simple model, incorporating three equations, for describing the transition in thermal convection to chaotic behavior. Lorenz (1963a) in a classic paper sought to understand forced dissipative flows and, especially, how deterministic descriptions could be "unstable with respect to small modifications, so that slightly differing initial states can evolve into considerably different states." He focused on the numerical solution of a simple system representing cellular convection manifesting intrinsic instability and sought to understand the implications of this work to long-range weather prediction. Thermal convection is often cited as a source of chaos (Drazin, 1992), and the *Lorenz model* employs a set of three ordinary differential equations as an approximation for thermal convection in a fluid layer heated from below (Turcotte, 1997). Sparrow (2005) provides an in-depth analysis of the Lorenz equations.

The Lorenz model is typically given (Strogatz, 1994) by

$$\dot{x} = \sigma(y - z)$$
$$\dot{y} = rx - y - xz$$
$$\dot{z} = xy - bz \tag{2.165}$$

where x, y, and z represent coefficients in a Fourier series associated with physical variables and where σ, r, and b are real-valued parameters that are related to the Prandtl number (the ratio of the viscous to thermal diffusion rates), the Rayleigh number (which relates buoyancy and viscosity in competition with momentum and thermal diffusion), and the aspect ratio (geometry) of convective cells. If we wish to think of these equations as defining a flow velocity $\boldsymbol{v} = (\dot{x}, \dot{y}, \dot{z})$ as in (4.39), we immediately observe that $\nabla \cdot \boldsymbol{v} = -\sigma - 1 - b < 0$, suggesting that the volume associated with any region in the flow will decrease at an exponential pace.

We immediately observe, after setting the left-hand side of each of these equations to zero, that there are three steady solutions for this problem:

1. $(0, 0, 0)$,
2. $(+\sqrt{b(r-1)}, +\sqrt{b(r-1)}, r-1)$, and
3. $(-\sqrt{b(r-1)}, -\sqrt{b(r-1)}, r-1)$.

We note that the third of these stationary points is a rotation of π radians about the z-axis in the x-y plane of the second, and we will confine our analysis to the first pair of steady solutions.

We linearize the variables by making use of the following substitutions:

$$x = X + \delta x$$
$$y = Y + \delta y$$
$$z = Z + \delta z, \tag{2.166}$$

where X, Y, and Z are selected to conform with one of the steady solutions. Performing the linearization, we then obtain

$$\delta \dot{x} = \sigma(\delta y - \delta x)$$
$$\delta \dot{y} = r \delta x - \delta y - X \delta z - Z \delta x$$
$$\delta \dot{z} = X \delta y + Y \delta x - b \delta z. \tag{2.167}$$

We begin by exploring the neighborhood of the origin $(0, 0, 0)$ and observe that

$$\begin{pmatrix} \delta \dot{x} \\ \delta \dot{y} \\ \delta \dot{z} \end{pmatrix} = \begin{pmatrix} -\sigma & \sigma & 0 \\ r & -1 & 0 \\ 0 & 0 & -b \end{pmatrix} \begin{pmatrix} \delta x \\ \delta y \\ \delta z \end{pmatrix}. \tag{2.168}$$

As before, we seek solutions of exponential form

$$\delta x(t) = \delta x_0 \exp(\nu t), \tag{2.169}$$

and obtain

$$0 = \begin{vmatrix} -\sigma - \nu & \sigma & 0 \\ r & -1 - \nu & 0 \\ 0 & 0 & -b - \nu \end{vmatrix}. \tag{2.170}$$

The solutions for ν, sometimes referred to as *eigenmodes*, include $\nu = -b$ and

$$\nu = \frac{-(\sigma + 1) \pm \sqrt{(\sigma + 1)^2 + 4\sigma(r - 1)}}{2}. \tag{2.171}$$

We observe that the real parts of all roots are negative if $r < 1$. But if $r > 1$, then one root is positive and this renders the solution for initial values near the origin unstable.

We now consider $Z = r - 1$ with $X = Y = \pm\sqrt{b(r - 1)}$:

$$\begin{pmatrix} \delta\dot{x} \\ \delta\dot{y} \\ \delta\dot{z} \end{pmatrix} = \begin{pmatrix} -\sigma & \sigma & 0 \\ r - Z & -1 & -X \\ X & X & -b \end{pmatrix} \begin{pmatrix} \delta x \\ \delta y \\ \delta z \end{pmatrix}. \tag{2.172}$$

We again perform the eigenmode (sometimes referred to as Fourier) analysis and obtain

$$0 = \begin{vmatrix} -\sigma - \nu & \sigma & 0 \\ 1 & -1 - \nu & -X \\ X & X & -b - \nu \end{vmatrix}. \tag{2.173}$$

We focus now on the positive $X = Y$ case; the remaining case ultimately yields the same result. The cubic equation that emerges is

$$0 = \nu^3 + (\sigma + 1 + b)\nu^2 + b(\sigma + r)\nu + 2\sigma b(r - 1). \tag{2.174}$$

When $r = 1$, we immediately observe that one root is zero and the other two have negative real parts, namely, $-\sigma - 1$ and $-b$. We can in principle employ the methods demonstrated in chapter 1 for finding the roots of a polynomial. The general methodology for identifying this bifurcation is known as the *Routh–Hurwitz theorem*; however, there is a simpler way.

A more practical approach emerges from identifying when the real part of at least one of the three roots vanishes, yielding a bifurcation in the system's behavior. We employ methods analogous to those we developed for solving cubic polynomials emerging from the analysis of symmetric 3×3 tensors to show, as r increases above 1, that the two roots having the largest but negative real parts are a complex-conjugate pair, and that the real parts of the two roots increase as r increases. Hence, we need only identify when that complex-conjugate pair becomes pure imaginary. We recognize that the three polynomial coefficients are the sum of the roots (with their signs reversed), the sum of all pairwise products of the roots (with their signs reversed), and the product of all roots (again with all signs changed). Suppose we assume that the real root is κ and the imaginary roots are $\pm\rho$. After a little algebra, we observe that the bifurcation occurs when the product of the ν^2 coefficient times the product of the ν^1 coefficient equals the ν^0 coefficient, namely,

$$r = \frac{\sigma(\sigma + 3 + b)}{\sigma - 1 - b}. \tag{2.175}$$

When r exceeds this value, two of the linearized damping modes develop oscillatory behavior, and, as r is systematically increased, the real parts of the eigenvalues ν become positive.

In order to demonstrate the nature of the chaotic flow, we have employed $r = 28$, $\sigma = 10$, and $b = 8/3$ in simulating the evolution of the Lorenz model shown in Figure 2.9.

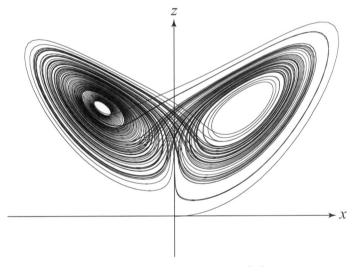

Figure 2.9. Lorenz attractor and chaos.

Owing to its unique character, plots of this sort have come to be called "butterfly diagrams" inasmuch as they expose the complex calligraphy of chaotic dynamics. It can readily be shown via computer experiments, albeit the analytic development is substantially more difficult, that performing an accurate numerical integration employing two different but nearby initial conditions will yield very different results. In order to characterize the exponential growth of the separation distance $d(t)$ between two solutions having slightly different initial conditions with initial separation distance $d(0)$, we introduce the *Lyapunov exponent* λ (Strogatz, 1994; Nayfeh and Balachandran, 1995) according to

$$\lambda = \lim_{t \to \infty} \frac{\ln[d(t)/d(0)]}{t}. \tag{2.176}$$

Some variations exist in the literature relating to this definition. In practice, it is not uncommon to see estimates of the Lyapunov exponent presented over a limited period of time.

2.4.4 Fractals

One intriguing feature of the butterfly diagram is the intricate character of the trajectories, which present a geometrical feature known as *self-similarity*. Imagine looking at part of the "wing," and then looking at it under a magnifying glass. What emerges is an approximately self-reproducing pattern on many different scales that is referred to as a *fractal* (Turcotte, 1997) and introduces a concept known as the "fractal dimension." We introduce in Figure 2.10 one of the simplest forms of fractal. We begin with a line segment, followed by a reproduction of it where a segment corresponding to its middle third has been eliminated.

Figure 2.10.   Cantor dust.

We repeat this procedure a number of times, and wish to develop a quantitative measure of its "dimensionality" D. It is clearly not $D = 1$ because the gaps ultimately guarantee that the cumulative "length" of the segments converges to zero. The index n shown in Figure 2.10 describes the level within this hierarchy, with $n = 0$ corresponding to a single segment length ℓ_0. We observe that the total length covered at that level L_0 is the number of segments N_n or 2^n multiplied by the length of the individual segment(s) $\ell_n = L_0/3^n$.

This kind of construction is referred to as *Cantor dust*, and has analogues in two and three dimensions. Importantly, the iterative procedure we employed here is completely regular and deterministic, and Cantor dust is a form of *deterministic fractal*. However, it need not be so: Rather, it could incorporate substantial randomness. For example, we could eliminate on each iteration a variable fraction of the length of each line segment centered at some random location. So-called *stochastic fractals* occur in many natural settings, particularly in geology and geophysics (Turcotte, 1997). The appearance of fractal structure in nonlinear systems has presented an important new advance in the study of nature and of mathematics.

Hence, the total length L_n at level n is simply

$$L_n = N_n \ell_n = 2^n 3^{-n} L_0. \tag{2.177}$$

How does this relate to dimensionality? Intuitively, we expect that the fractal dimension D is related to the number of segments according to

$$N_n \propto \ell_n^{-D}. \tag{2.178}$$

We eliminate the proportionality constant to isolate the n-dependence and observe that

$$2^{-n} = 3^{-nD}, \tag{2.179}$$

giving

$$D = \frac{\ln 2}{\ln 3} \approx 0.6309298. \tag{2.180}$$

While the notion of a noninteger dimension may run contrary to simple intuition, it should be regarded as displaying that complex structures do not abide by simple geometric rules. An idealized mathematical surface is two-dimensional. Real surfaces, as investigated in tribology, however, show that self-similarity often abides and that the "fuzziness" associated with a physical surface is an indication that its fractal dimension is greater than 2, albeit smaller than 3.

To illustrate this, we show in Figure 2.11 a *Koch snowflake* (Mandelbrot, 1983; Peitgen et al., 1988; Feder, 1988; Bak, 1996; Turcotte, 1997). As before, we begin with a line segment and remove the middle third, only to replace it with two equal-length line segments as though we were completing an inverted equilateral triangle. We then iterate, replacing each line segment by

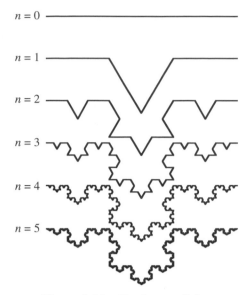

$n = 0$

$n = 1$

$n = 2$

$n = 3$

$n = 4$

$n = 5$

Figure 2.11. Koch snowflake.

four. The end result appears "fuzzy," as mentioned, and has a fractal dimension of $\ln 4 / \ln 3 \approx 1.2618595$, twice the value observed for the Cantor dust. It is clear that the fractal dimension is slightly greater than 1.

While these ideas are a departure from conventional thinking, they are an integral part of geophysics. In an often cited story (Mandelbrot, 1983), the prominent applied mathematician and hydrodynamicist Lewis Fry Richardson was asked "What is the length of the coastline of Britain?" to which he responded, "That depends upon the length of the ruler that I use." The analogy with the methodology used above to generate Cantor dust and Koch snowflakes is clear; the "length" of the structure depends critically upon the size of the ruler used, that is, the resolution present. These methods can be extended to generating realistic-looking "natural" surfaces (Peitgen et al., 1988). Feder (1988), Turcotte and Newman (1996), and Turcotte (1997) have described how fractals play a fundamental role in the earth sciences, and considerable speculation has emerged on how such "scaling" could apply to other problems, such as the Gutenberg–Richter power-law relation for earthquakes (Newman, 2012), which describes the frequency of seismic events as a function of their magnitude. This is also related to the Kolmogorov spectrum for atmospheric turbulence that we derive at the end of chapter 4. Bak (1996) has proposed that many microscopic processes could cascade into power-law behaviors via a mechanism he calls "self-organized criticality." This arena remains a vigorous one for research in geophysics and many other sciences and engineering.

We have now observed that relatively simple mathematical models, such as that due to Lorenz (1963a), can manifest incredibly intricate and complicated behavior. Using the Bullard (1955) homopolar dynamo model, we reconstructed early efforts in geophysics to describe quasiperiodic phenomena emerging from the interaction between the magnetic field generated by the Earth's dynamo and the rotational motion associated with our planet's spin, only to note that phenomena such as field reversals, intermittency, and so on were not within reach. In biology, similar efforts to understand cycles evident in the population of mammals through *predator–prey* models coupling ordinary differential equations governing the interaction between the two species encountered similar problems. The Poincaré–Bendixson theorem provided a fundamental barrier to those seeking to model quasiperiodic behavior in terrestrial dynamo and mammalian population dynamics. Remarkably, while only slightly more complicated, the Lorenz model (Sparrow, 2005) presents a dramatic reduction in the physical richness of the underlying partial differential equations describing convection and the pathway to turbulent flow—although it was not meant to describe convection. Its development demonstrated that simple ordinary differential equation models could capture the essence of complicated systems in three or higher dimensions.

May (1976) demonstrated, in a classic paper dedicated to problems emergent in biology, economics, and the social sciences, that even simpler models—using in essence finite difference equations or mappings—could yield very complicated dynamics. May's focus was on the generational change in a population x of some species, and he employed a simple mapping equation for generation $n + 1$ in terms of generation n, which we will express as

$$x_{n+1} = g(x_n), \tag{2.181}$$

where the function $g(x)$ provides a description of how the population in one generation begets the population of the next. How to define this function is not immediately obvious. One could fit observational (field) data, for example, or one could fit approximate solutions to the underlying differential equations.

2.4.5 Maps and Period Doubling

May (1976) employed the latter approach, beginning with the *logistic equation* developed by Verhulst more than a century earlier, for the population ρ of a species relative to the carrying capacity of the environment, which is usually given as

$$\frac{d\rho(t)}{dt} = \gamma\rho(t)[1 - \rho(t)]. \qquad (2.182)$$

For small values of $\rho \ll 1$, far below the carrying capacity of the environment, we see essentially exponential or *Malthusian* growth. However, as ρ increases, saturation effects take over. Remarkably, if we define

$$v = \frac{\rho}{\beta} - v_t, \qquad (2.183)$$

we recover Eq. (2.83) for Rayleigh drag, for which we derived in the text and exercises three equivalent methods of solution showing exponentially rapid convergence to the fixed point. In the population dynamic case, this corresponds to the carrying capacity of the environment, which we have scaled to 1 here. See Hoppensteadt (1982) and Murray (2003) for a detailed discussion of the biological aspects of the latter equation. Importantly, for any initial condition $\rho(0) > 0$, the long-time solution is always 1 and, in the neighborhood of the carrying capacity, this problem is remarkably stable.

Using a commonly used argument (May, 1976; Feigenbaum, 1980), let us explore the behavior of (2.182) if it were approached using a very simple numerical method due originally to Euler, which we will discuss further in chapter 5. We "sample" the population, as though we were performing a census, at regular intervals $t_n = t_0 + n\Delta t$ corresponding to a generational change to obtain $\rho(t_n) = \rho_n$. Further, we replace the time derivative in (2.182) by its approximate value

$$\frac{d\rho(t)}{dt} \approx \frac{\rho_{n+1} - \rho_n}{\Delta t}, \qquad (2.184)$$

whereupon (2.182) can be replaced by

$$\rho_{n+1} = (1 + \gamma\Delta t)\rho_n\left[1 - \frac{\gamma\Delta t}{1 + \gamma\Delta t} \cdot \rho_n\right]. \qquad (2.185)$$

Employing the substitution

$$x_n = \frac{\gamma \Delta t}{1 + \gamma \Delta t} \rho_n, \tag{2.186}$$

we obtain the standard form for what is called the logistic or quadratic or Feigenbaum map:

$$x_{n+1} = \alpha x_n (1 - x_n), \tag{2.187}$$

where

$$\alpha = 1 + \gamma \Delta t. \tag{2.188}$$

Another frequently employed variant of this (Feigenbaum, 1980) replaces x_n by $y_n + 1/2$ to get $y_{n+1} = \alpha(1/4 - y_n^2)$. It should be clear that (2.187) is quantitatively accurate when $\gamma \Delta t \ll 1$, that is, so that the time step $\gamma \Delta t$ is much smaller than the exponential growth time γ^{-1}. Let us now explore some features of the general map (2.181).

Let ζ be a solution to (2.181), that is,

$$\zeta = g(\zeta). \tag{2.189}$$

Subtracting this from (2.181) and using a Taylor series, we observe that

$$|x_{n+1} - \zeta| = |g(x_n) - g(\zeta)| \approx |x_n - \zeta||g'(\zeta)|. \tag{2.190}$$

Therefore, it follows that

$$\frac{|x_{n+1} - \zeta|}{|x_n - \zeta|} \approx |g'(\zeta)|, \tag{2.191}$$

and we observe that a convergence criterion for the sequence x_n is that

$$|g'(\zeta)| < 1. \tag{2.192}$$

We note in order for the map (2.187) to produce non-negative x_{n+1} that $0 < x_n < 1$. Further, since

$$x_n(1 - x_n) = (\tfrac{1}{2})^2 - (x_n - \tfrac{1}{2})^2, \tag{2.193}$$

the largest value that x_{n+1} can obtain is $\alpha/4$, thereby requiring that $\alpha \leqslant 4$.

We observe that the map (2.187) has two fixed points ζ. The first, $\zeta = 0$, presents

$$|g'(0)| = \alpha = 1 + \gamma \Delta t > 1. \tag{2.194}$$

Consequently, this fixed point is always unstable, however small we make Δt. This is analogous to the behavior present in the ordinary differential equation (2.182) that manifests exponential growth for ρ near 0. The second fixed point satisfies

$$\zeta = 1 - \frac{1}{\alpha} = \frac{\gamma \Delta t}{1 + \gamma \Delta t}, \tag{2.195}$$

and thereby corresponds to $\rho = 1$, the carrying capacity of that population. If this fixed point is stable, then that corresponds directly to the solution of the underlying ordinary differential equation (2.182). We observe that

$$\left| g'\left(1 - \frac{1}{\alpha}\right) \right| = |2 - \alpha|. \tag{2.196}$$

Hence, the map is stable in the vicinity of the fixed point only when $1 < \alpha < 3$, where we have already guaranteed that the lower bound is satisfied.

We show in Figure 2.12 the case when $\alpha = 2.7$, which satisfies the upper bound for stability just derived.

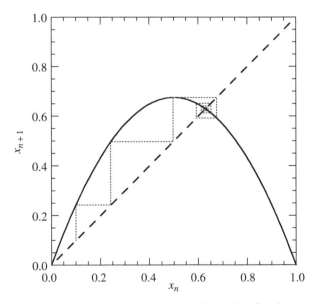

Figure 2.12. Quadratic mapping with stable fixed point.

The plot describes the relationship between x_{n+1} and x_n for (2.187). The map itself is shown via a solid line depicting the downward pointing parabola. The dashed line at 45° describes the equation $x_{n+1} = x_n$. We initiate our calculation with $x_0 = 0.1$ and show the point $(0.1, 0.1)$ on the lower right of the diagonal dashed line. That initial point rises according to the map to the parabola with the calculation to x_1 and then assumes its position on the dashed line as the new value of the population. It rises again to the parabola, moves to the dashed line, and so on, as the iteration of the mapping proceeds. After awhile, when the sign of $g'(x_n)$ changes but $|g'(x_n)|$ remains less than unity, the trajectory of the mapping assumes a pattern that oscillates and closes in on the fixed point. The stability observed here loosely mimics that underlying the ordinary differential equation (2.182) but is more complicated.

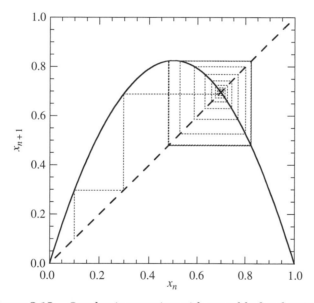

Figure 2.13. Quadratic mapping with unstable fixed point.

In Figure 2.13, we bring α above the stability limit to a value of 3.3, and initiate our calculation in the same way. The trajectory initially behaves in the same way but, once it becomes close to the new fixed point, responds to it being unstable and, instead of "spiraling" inwards towards it, spirals out. However, we observe a remarkable feature of its evolution, reminiscent of limit cycles in the Van der Pol oscillator: The spiral becomes self-limiting, and we observe an oscillation between what appears to be two fixed points.

In Figure 2.14, we describe the geometry of the composite mapping

$$g^{(2)}(x) \equiv g[g(x)], \tag{2.197}$$

where we have employed the superscript index (2) to indicate that g has acted upon itself again for $\alpha = 3.3$.

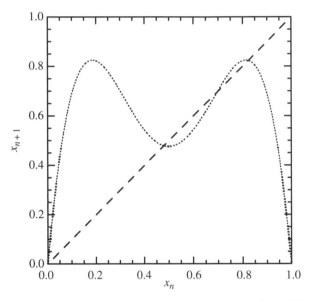

Figure 2.14.   Quadratic mapping showing period doubling.

We observe three intersections of $g^{(2)}(x)$ with the diagonal line. The intermediate of these two is the unstable fixed point of $g(x)$ that we observed earlier. However, we see that there are two further fixed points for $g^{(2)}(x)$ where the slopes at the intersection points are less than 1 in magnitude. Thus, we observe that the two new fixed points of $g^{(2)}(x)$ are stable and that the mapping $g(x)$ provides a periodicity of 2. We refer to this qualitative change in behavior as presenting a *period-doubling bifurcation*. With further increases in the parameter α, repeated period-doubling bifurcations occur, that is, we will present a sequence of mappings $g^{(4)}(x)$, $g^{(8)}(x)$, ..., $g^{(2^m)}(x)$, ..., for $m = 4, 5, \ldots$. When we plot the behavior of the fixed points as a function of α (Figure 2.15), we would observe a sequence of *pitchforks* wherein a smooth function of α, namely, the fixed point, would split into two curves, and then into four, and so on.

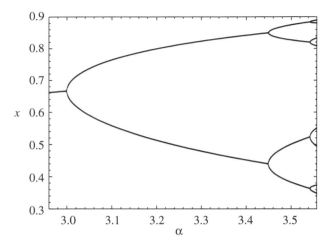

Figure 2.15. Period-doubling bifurcation sequence; note that the pattern of doublings becomes much more intricate and complex as α increases further.

However, this pattern breaks down beyond $\alpha \approx 3.56$. For many years, it was widely believed that the pitchfork bifurcation resided at the heart of all routes to chaos. The period-doubling phenomenon in mappings as well as in nature is described in Moon (1992), Hilborn (2000), and many other books. Mathematically, as elegantly described by Feigenbaum (1980), certain features of the period-doubling phenomenon are universal and underscore a very broad class of maps $g(x)$. Indeed, universal scaling features of these bifurcations and their dependence upon α and the emergence of universal numbers relating to their scaling—often called Feigenbaum numbers in honor of their discoverer—have had a profound influence upon our thinking about nonlinear phenomena.

Before advancing to our next topic, let us now explore the limiting case $\alpha = 4$, which evidently was done by Von Neumann during the early days of digital computers. In addition to this selection of α, we also make the substitution

$$x_n = \sin^2(\theta_n), \tag{2.198}$$

whereupon we observe that the new variable θ_n satisfies the recurrence relation

$$\theta_{n+1} = 2\theta_n. \tag{2.199}$$

Since the sine function is periodic with period 2π, we refer to (2.199) as the *circle map*. Within the constraints posed by the finite accuracy of digital computers, this mapping enjoyed a brief history as a device for generating *pseudorandom numbers*. One would begin with an initial value or "seed" θ_0 and repeatedly double it modulo 2π. Of course, this could only be done a limited number of times because, mathematically, this corresponded to shifting a floating-point number in computer memory by one bit position to the left. Nevertheless, ideas of this sort had a profound role in the evolution of computer science and the later emergence of Monte Carlo simulations in many arenas of exploration. We will discuss this issue further in chapter 5.

While this is of mathematical interest, does this phenomenon have any relevance to the physical world? In planetary science, the rotation of Saturn's satellite Hyperion (Murray, 2003) and the orbits of the Uranian satellites (Malhotra et al., 1989) exhibit chaotic orbits. Real systems present many different routes to chaos, with period doubling among them. In Rayleigh-Bénard convection, the phenomenon underlying the work performed by Lorenz, *observations* of bifurcations from 1 to 2 to 4 cells and so on present some of the earmarks of this mathematical phenomenon. Moon (1992) provides an overview of many of these phenomena, while the semipopular book by Gleick (2008) provides a relatively nontechnical survey of what we have learned in this domain. Indeed, issues emerging from systems including large numbers of variables or "degrees of freedom" have transformed the elementary ideas that we have discussed to the richer context of *complexity*. Some aspects of this phenomenon will be discussed in chapter 4.

We depart now from our treatment of ordinary differential equations—and our momentary diversion to fractal structure, mappings, and chaos—but will return to ordinary differential equations when we explore the special functions that emerge from second-order ordinary differential equations in a variety of special applications and en route to developing a methodology for obtaining the solutions to the partial differential equations of mathematical geophysics identified in this chapter. We now turn to the evaluation of integrals by a variety of methods, including complex analysis and contour integration, as well as explore some special integrals, Fourier methods, and inverse theory encountered in geophysics.

2.5 Exercises

1. One of the earliest methods developed for addressing ordinary differential equations emerges from finding *power series solutions* (Jeffreys and Jeffreys, 1999). Consider as an example finding the solution to

$$\ddot{x}(t) + \omega^2 x(t) = 0,$$

   subject to $x(0) = 0$ and $\dot{x}(0) = 1$. Suppose that we express the power series as

$$x(t) = \sum_{n=0}^{\infty} a_n t^n,$$

   where the a_n are the power series coefficients to be determined.

   (a) Using the initial conditions, find a_0 and a_1.

   (b) Employing the defining differential equation, find the recurrence relation relating a_{n+2} with a_n, for $n = 0, 1, \ldots$.

   (c) Provide an explicit formula for the a_n and, thereby, show how this solution is equivalent to a certain trigonometric function.

2. Suppose you have a Newtonian force whose potential is

$$V(x) = -\tfrac{1}{2}kx^2,$$

   differing by a sign from the usual simple harmonic oscillator. You may continue to call $k/m = \omega^2$, but note that ω is no longer an oscillator frequency.

   (a) What is the homogeneous second-order differential equation that must be solved?

   (b) What are the two homogeneous solutions that emerge for it?

3. Continue with this repulsive Newtonian force problem.

   (a) Why does one of these become unimportant at late time? (This underscores a property of second-order differential equations over a domain. It is generally the case one solution is "regular" and one is "irregular" over the domain. We will observe later the significance of this observation.)

(b) Sketch the phase portrait for this problem. What kind of curve have you drawn? [Hint: It is no longer a circle or ellipse.]

4. Derive the equation for Q shown in (2.57).

5. Finding two homogeneous solutions, one regular and the other irregular, of a linear second-order differential equation such as

$$a(t)\ddot{x}(t) + b(t)\dot{x}(t) + c(t)x(t) = 0$$

is not always easy.

(a) Suppose that $f(t)$ is a solution to the preceding equation. We can construct a second solution by making the assumption that it can be written $f(t)g(t)$. What first-order differential equation must $g(t)$ satisfy?

(b) Solve that equation for $g(t)$.

6. Consider again the LRC equation (2.53), where we will assume that the forcing voltage $V(t) = A \sin \Omega t$ where $\Omega < |\omega_{\pm}|$, that is, the forcing frequency is smaller than the natural frequency of the problem, that is, the real part of $\omega_{\pm}$. Show that $q(t)$ will have the form $B \sin(\Omega t + \theta)$ and that $\theta < 0$—the current shows a "phase lag" with respect to the driver.

7. Suppose, due to a nonlinear forcing, that you must address a differential equation of the form

$$\ddot{x}(t) + \omega^2 x(t) = A \cos^2 \Omega t.$$

(a) Can a resonance occur for this problem? If so, for what frequency? Derive your results.

(b) What is the general solution to this problem? You may use any of the methods described in the text to derive this.

8. *Airy's equation* (Jeffreys and Jeffreys, 1999)

$$\frac{d^2 y(x)}{dx^2} = x y(x)$$

occupies an important role in geophysics, especially in seismology. Using the JWKB method, find an approximate solution to this equation—be sure to simplify the integrals that emerge. What issues emerge if we sought to extend the solution to $x < 0$? This is sometimes referred to as *Stokes's phenomenon* (Kemble, 2005).

9. Let us return to the problem of Rayleigh atmospheric drag with terminal velocity (2.83).

   (a) We mentioned earlier that Riccati equations can be converted into second-order linear differential equations. Making the substitution

   $$v = \frac{1}{\alpha} \frac{\dot{w}}{w},$$

   show that conversion to second-order is possible and solve the ensuing linear equation.

   (b) The original equation (2.83) is separable. Show that

   $$\left[\frac{1}{v_t - v} + \frac{1}{v_t + v} \right] \dot{v} = 2\alpha v_t.$$

   Then, show that

   $$\frac{d}{dt} \left[\ln \left(\frac{v_t + v}{v_t - v} \right) \right] \dot{v} = 2\alpha v_t.$$

   Finally, using u in place of the quantity above in square brackets, that is, the logarithmic term, derive the final result, where c is a constant, that

   $$v(t) = v_t \left\{ \frac{\exp[2\alpha v_t t + c] - 1}{\exp[2\alpha v_t t + c] + 1} \right\}.$$

10. There are many situations, including crystallography and mineralogy, where equations of the form

    $$\frac{d^2 y(x)}{dx^2} + k^2(x) y(x) = 0$$

    are encountered where $k^2(x)$ is periodic, that is,

    $$k^2(x + L) = k^2(x).$$

    How does this periodicity affect the solution for $y(x)$? There is a substantial literature on what has come to be called *Floquet's theorem* and *Mathieu functions*, which are a special case. Using JWKB analysis, what can you say about $y(x + L)$ in terms of $y(x)$? [Hint: You may want to consult a more exhaustive treatment of mathematical methods to see what is Floquet's theorem.]

11. Consider the periodic motion present in a pendulum whose angular frequency

$$\omega^2 = \frac{g}{\ell},$$

where g is the local gravitational acceleration and ℓ is the length of the string. Suppose, now, that the string supporting the pendulum is held in your hand, and is gradually shortened by pulling on it so that its length is a slowly varying function of time $\ell(t)$. This problem can be treated using Floquet's theorem. Perform a JWKB analysis for this problem to obtain an approximate solution for this problem. What is the adiabatic invariant for this problem, and how does it relate the instantaneous energy in the pendulum motion to the pendulum's length ℓ?

12. Consider a box with sides L_x, L_y, and L_z where a standing-wave amplitude Ψ vanishes on the boundaries (box sides). Identify the eigenvalue conditions, and determine the solution to Helmholtz's equation.

13. Prove that the series expansion (2.111) satisfies Bessel's equation (2.110).

14. Using the first of the Bessel function generators (2.113), show that

$$J_{n+1}(r) = \frac{2n}{r} J_n(r) - J_{n-1}(r).$$

[Hint: Take the derivative with respect to t of the generating function.]

15. Similarly, by taking the derivative with respect to r, show that

$$J_n'(r) = \tfrac{1}{2}[J_{n-1}(r) - J_{n+1}(r)].$$

16. Suppose you have a drum of radius R and you wish to establish a standing wave, that is, a wave satisfying the Helmholtz equation where Ψ is the amplitude of the oscillation of the drum.

(a) Given that the edge of the circular drum is fixed, what is the boundary condition that this establishes on the oscillation?

(b) For a given value of k, are oscillations at all possible, or is this an eigenvalue problem? If the latter is the case, identify the condition that a circularly symmetry oscillation must satisfy. [Hint: It will involve both k and R as well as properties of $J_n(r)$ for a special value of n.]

(c) What can you say for oscillations that are twice periodic, that is, involve $\exp(\pm i 2\theta)$?

17. Consider now a cylinder with height H and radius R with a standing wave present inside such that the wave amplitude Ψ satisfies Helmholtz's equation. Identify the eigenvalue conditions and determine the solution.

18. For integer ℓ, prove that Rodrigues's formula (2.136) satisfies the Legendre equation (2.135).

19. Again using Rodrigues's formula (2.136), prove the orthogonality relation (2.138).

20. Using the Legendre generating function (2.136), derive the recurrence relation

$$(\ell + 1)P_{\ell+1}(x) = (2\ell + 1)xP_\ell(x) - \ell P_{\ell-1}(x).$$

21. Similarly, prove

$$P'_{\ell+1}(x) = xP'_\ell(x) + (\ell + 1)P_\ell(x).$$

22. Compare the definitions employed for spherical harmonics presented here, as well as in Jackson (1999), with the two definitions provided by Stacey and Davis (2008). Develop a table describing all three systems, and provide a formula that will allow you to navigate from any one of these systems to the other two.

23. Show that
$$\nabla^2 Y_{\ell,m} = -\frac{\ell(\ell + 1)}{r^2} Y_{\ell,m}.$$

24. The gravitational potential of the Earth, $\Phi(r, \theta, \phi)$, can be related to the mass density $\rho(r, \theta, \phi)$ according to

$$\nabla^2 \Phi(r, \theta, \phi) = 4\pi G \rho(r, \theta, \phi).$$

Expanding both the potential and the mass density in spherical harmonics, find the differential equation in r relating $\Phi_{\ell n}(r)$ to $\rho_{\ell n}(r)$.

25. Consider, as did Bullard, the homopolar dynamo problem (2.155) and (2.156). Following Bullard, show that

$$\frac{d^2 \ln I}{dt^2} = \frac{M}{2\pi L} \frac{d\omega}{dt}.$$

Using (2.156) and defining $y = \ln I$, show that there is a quadrature governing the time evolution with a constant C'' defined by

$$C'' = \frac{1}{2}\left(\frac{dy}{dt}\right)^2 - \left[\frac{G}{2\pi CL}y - \frac{M^2}{8\pi^2 CL}\exp(2y)\right].$$

26. Return to problem (3.2) and the definition of the Lyapunov exponent (2.176). For convenience, we will identify $y(t)$ with $\dot{x}(t)$. Derive the Lyapunov exponent for this problem assuming that the "distance" between solutions is simply $\sqrt{x^2 + y^2}$.

27. Consider the driven Van der Pol oscillator problem

$$\ddot{x}(t) - \mu[1 - x^2(t)]\dot{x}(t) + x(t) = f(t),$$

with $\mu > 0$.

   (a) Use perturbation theory to identify the lowest order behavior of this system, and solve the associated equations. The method of variation of constants is especially valuable.

   (b) Suppose the forcing function was $\sin(t)$. Ignoring, for the moment, the nonlinearity in the problem, will resonant behavior and a secular solution be a problem? Why or why not?

   (c) For a general forcing function, not simply the sinusoid used in the previous problem segment, does the Poincaré–Bendixson theorem offer any insight into the overall behavior of this system? Why or why not? [Hint: Consider the conditions associated with the trajectory in the original autonomous problem to not cross over themselves.]

28. The two-dimensional analogue of Cantor dust is referred to as a *Sierpinski carpet* (Turcotte, 1997). Beginning with a square, subdivide it into 9 equal-sized squares, eliminating in each direction the middle third, that is, retain 4 subsquares out of the original 9. What is the fractal dimension of this object? In three dimensions, employing a similar procedure, it is possible to construct a *Menger sponge* where a 27-element cube retains only 8 components under an iteration of this scheme. What is its fractal dimension?

29. Consider the construction of the Koch snowflake presented earlier, in contrast with Cantor dust, and derive its fractal dimension.

CHAPTER THREE

Evaluation of Integrals and Integral Transform Methods

Integrals provide an equivalent approach to solving first-order ordinary differential equations (2.3). It is quite common in geophysical applications to encounter problems where it becomes necessary to evaluate an integral. Often, these integrals involve exponentials and special methods or techniques are invaluable in evaluating these, for example, integrals involving Gaussian distributions or producing factorials (gamma function). In other instances, those same methods can help provide estimates for integrals involving exponentials of functional quantities that effectively isolate only a small region of the domain of integration and so-called methods of steepest descent or saddle point methods become invaluable. Texts, such as those by Mathews and Walker (1970) and by Arfken and Weber (2005) provide more insight into these issues. There also exist a modest number of special integrals that repeatedly appear in a variety of problem areas, for example, elliptic integrals. Jeffreys and Jeffreys (1999) as well as Whittaker and Watson (1979) are especially valuable for these topics. We will provide a brief survey of each of these problem categories.

The application of complex analysis and the calculus of residues makes it possible via contour integration to solve many classic problems. Textbooks, such as that by Churchill (1960), provide a much more comprehensive introduction to the subject. We will briefly review and survey these methodologies, as they become invaluable in solving linear partial differential equations.

The term *integral transform* in mathematical methods courses is usually employed to denote Fourier series and integrals, and we will review these here, mentioning in passing the Laplace transform and the Bromwich contour. In geophysical applications, the issue of convolution and, especially, deconvolution becomes very important, so we will also undertake a survey of some of these issues, exploring both deterministic and stochastic problems. These issues are also important in *geophysical inverse theory* (Parker, 1977, 1994). Theoretical treatments generally overlook issues pertinent to practical considerations, such as sampling, and the nature of approximations. We will derive the sampling theorem and identify the meaning of "aliasing" so that the application of the methods described here will not introduce unnecessary error. Meanwhile, we will review some of the approximations that emerge in this context, and how discrete Fourier transforms can be employed and, especially, be dramatically accelerated without introducing further approximations through the fast Fourier transform. Integral equations occur frequently in geophysics, so their inversion becomes especially important and establishes the basis of geophysical inverse theory. To this end, we review the Abel transform and its generalization to problems lacking rotational symmetry, that is, the Radon transform. This takes us into the realm of computed tomography and, finally, the Herglotz–Wiechert transform widely employed in seismology.

For more incisive treatments of these problems, Bracewell (2000) presents a very nice introduction to the Fourier transform and its applications. Walker (1988) provides a mathematically rigorous treatment of these and many other topics. Box et al. (2008) pursue questions attendant to stochastic, in contrast with deterministic, processes and the methodology that becomes necessary, as does the classic Blackman and Tukey (1958). Brigham (1988) and Nussbaumer (1982) present comprehensive overviews of fast Fourier transforms. Finally, Deans (2007), as well as Walker (1988) mentioned earlier, provide detailed treatment of the Radon transformation.

3.1 Integration Methods, Approximations, and Special Cases

We begin this chapter, by exploring a variety of integration methods, including asymptotic approaches and other approximations, steepest descent methods, and special integral forms that occur in geophysics, particularly elliptic integrals.

3.1.1 Elementary Methods and Asymptotic Methods

In this section, we will be encountering a variety of integral types that we wish to evaluate, primarily ones involving exponentials.

We will identify these different cases using the Greek letter Ξ together with a subscript that identifies the case being investigated and, if necessary, will present in parentheses the parameter (or parameters) incorporated into its definition.

We begin with the integral associated with the Gaussian or normal distribution, which we will write for $\alpha > 0$ as

$$\Xi_1(\alpha) \equiv \int_{-\infty}^{\infty} \exp(-\alpha x^2) \, dx. \tag{3.1}$$

The standard method for evaluating this integral is to take its square and use both x and y integration variables, namely,

$$\Xi_1^2(\alpha) = \int_{-\infty}^{\infty} \exp(-\alpha x^2) \exp(-\alpha y^2) \, dx \, dy$$

$$= \int_0^{\infty} \int_0^{2\pi} \exp(-\alpha r^2) r \, dr \, d\theta, \tag{3.2}$$

where we have switched to polar coordinates. The θ integration yields a factor of 2π, and replacing αr^2 by z, we observe that

$$\Xi_1^2(\alpha) = 2\pi \int_0^{\infty} \exp(-z) \frac{1}{2\alpha} \, dz = \frac{\pi}{\alpha}, \tag{3.3}$$

and

$$\Xi_1(\alpha) = \sqrt{\frac{\pi}{\alpha}}. \tag{3.4}$$

As an illustration of the use of this result, we observe that a normalized Gaussian distribution $\mathcal{N}(x; \mu, \sigma)$ with a mean or first moment μ and variance or second moment σ^2—more on this topic in chapter 5—has the form

$$\mathcal{N}(x; \mu, \sigma) = \frac{1}{\sqrt{2\pi\sigma^2}} \exp\left[-\frac{(x-\mu)^2}{2\sigma^2} \right] \tag{3.5}$$

and satisfies

$$1 = \int_{-\infty}^{\infty} \mathcal{N}(x; \mu, \sigma) \, dx. \tag{3.6}$$

Since $\mathcal{N}$ is symmetric around μ, it follows directly that

$$\mu = \int_{-\infty}^{\infty} x \mathcal{N}(x; \mu, \sigma) \, dx. \tag{3.7}$$

Finally, note that we can evaluate by similar means related integrals such as

$$\Xi_2(\alpha) \equiv \int_{-\infty}^{\infty} x^2 \exp(-\alpha x^2) \, dx = -\frac{d}{d\alpha}[\Xi_1(\alpha)] = \frac{1}{2\alpha}\sqrt{\frac{\pi}{\alpha}}, \tag{3.8}$$

and, with a little algebra, we obtain

$$\sigma^2 = \int_{-\infty}^{\infty} (x - \mu)^2 \mathcal{N}(x; \mu, \sigma)\, dx. \tag{3.9}$$

In like manner, we can obtain all (integer) moments of $\mathcal{N}$.

Another commonly encountered integral involving exponentials is related to the gamma function $\Gamma(z)$, for $z > 0$,

$$\Gamma(z) \equiv \int_0^{\infty} \exp(-x) x^{z-1}\, dx. \tag{3.10}$$

We directly observe that $\Gamma(1) = 1$, and, integrating by parts, we can verify that

$$\Gamma(z + 1) = z\Gamma(z), \tag{3.11}$$

showing, for non-negative integer z, that

$$\Gamma(z + 1) = z! \tag{3.12}$$

establishing the relationship between the gamma function and the factorial, for example, $\Gamma(2) = 1!$; note that the gamma function will have poles for $z = 0, -1, -2, \ldots$. It is natural to ask, then, what is $\Gamma(3/2)$, namely,

$$\Gamma\left(\frac{3}{2}\right) = \int_0^{\infty} \exp(-x) x^{1/2}\, dx. \tag{3.13}$$

We make the substitution $x = y^2$ and obtain an expression closely related to our second moment identity; hence, we observe that

$$\Gamma\left(\frac{3}{2}\right) = \frac{\sqrt{\pi}}{2}. \tag{3.14}$$

There are many other problems and results emergent from the gamma function; Whittaker and Watson (1979) and Jeffreys and Jeffreys (1999) are excellent resources on these topics.

Our focus in this section has been on integrals evaluated over an infinite domain. For example, consider $\Xi_3(x)$ defined by

$$\Xi_3(x) \equiv \int_0^x \exp(-x'^2)\, dx', \tag{3.15}$$

which, in the limit $x \to \infty$, yields $\sqrt{\pi}/2$. For small values of x, we expand the exponential as a power series to obtain

$$\begin{aligned}
\Xi_3(x) &= \int_0^{\infty} \left[1 - \frac{x'^2}{1!} + \frac{x'^4}{2!} - \frac{x'^6}{3!} + \cdots \right] dx' \\
&= \left[x - \frac{x^3}{3 \cdot 1!} + \frac{x^5}{5 \cdot 2!} - \frac{x^7}{7 \cdot 3!} + \cdots \right].
\end{aligned} \tag{3.16}$$

It can be shown (Cauchy's test) that the last term calculated in this series provides an upper bound to the error in this approximation. The case of large x is treated differently through the calculation of an *asymptotic series* in the variable $1/x$. Unlike ordinary power series, asymptotic series are not formally convergent. These series tend to provide a sequence of smaller terms that then give way to increasing terms. However, in practice, we terminate the series when its terms cease to decrease and calculate an error bound. We will use $\Xi_3(x)$ for large x as an illustration.

We begin by rewriting (3.15) as

$$\Xi_3(x) = \frac{\sqrt{\pi}}{2} - \int_x^\infty \exp(-x'^2)\,dx'. \qquad (3.17)$$

The key here is to exploit "integration by parts" to convert the integral from x to ∞ to one where recursion can be exploited. We observe that

$$\int_x^\infty \exp(-x'^2)\,dx' = \int_x^\infty \frac{1}{2x'} \cdot \exp(-x'^2)2x'\,dx', \qquad (3.18)$$

and note that

$$\exp(-x'^2)2x'dx' = -d[\exp(-x'^2)]. \qquad (3.19)$$

Hence,

$$\begin{aligned}
\int_x^\infty \exp(-x'^2)\,dx' &= \left[-\frac{\exp(-x'^2)}{2x'}\right]_x^\infty - \int_x^\infty \frac{\exp(-x'^2)}{2x'^2}\,dx' \\
&= \frac{\exp(-x^2)}{2x} - \frac{1}{4}\int_x^\infty \frac{\exp(-x'^2)}{x'^3}2x'\,dx',
\end{aligned}$$
$$(3.20)$$

where we now employ integration by parts on the last term, recursively. While it may appear at this stage that the integer terms that appear in the integrals steadily decrease, the need to take successively higher derivatives of x^{-1} results in terms in the numerator having the form $3 \cdot 5 \cdot 7 \cdots$, which leads to the divergence in terms. However, we note that the remainder integrals can be bounded very effectively. For example, we note that

$$\int_x^\infty \frac{\exp(-x'^2)}{x'^3}2x'\,dx' \leqslant \frac{1}{x^3}\int_x^\infty \exp(-x'^2)2x'\,dx' = \frac{1}{x^3}. \quad (3.21)$$

Thus, while the series is not convergent, for very large x, we can calculate a large number of terms before they begin to increase and estimate the error in our asymptotic approximation. Now, we turn to another topic emergent from performing integrals over exponentials.

3.1.2 Steepest Descent Methods

There are a broad array of methods developed for evaluating integrals of the form

$$\Xi_4 = \int_{-\infty}^{\infty} f(x) \exp[-g(x)]\, dx, \qquad (3.22)$$

where $g(x)$ has a pronounced minimum at x_0 and can be approximated in that neighborhood as a quadratic, and these approaches are sometimes referred to as the saddle point method.

First, given our earlier introduction to the Dirac δ function, it follows that a possible representation for the former can be written as

$$\delta(x) = \lim_{\sigma \to 0} \frac{1}{\sqrt{2\pi\sigma^2}} \exp\left[-\frac{x^2}{2\sigma^2}\right]. \qquad (3.23)$$

We observe, in this limit, that the right-hand side very quickly goes to zero when x is nonzero, becomes singular at the origin, and has an integral of one. We will employ this result in a number of applications in later chapters.

Let us turn our attention to (3.22), where we assume that $g'(x_0)$ vanishes, so that we can write as a Taylor series

$$g(x) \approx g(x_0) + \frac{1}{2}\left[\frac{d^2 g(x)}{dx^2}\right]_{x=x_0} (x - x_0)^2 + \cdots . \qquad (3.24)$$

We have been somewhat imprecise with the error term here; fundamentally, we are presuming that higher-order terms contribute to making g much larger far from x_0. Accordingly, noting the δ function–like role of the exponential term and our previous results, we can write

$$\Xi_4 = f(x_0) \exp[-g(x_0)]\sqrt{\frac{2\pi}{g''(x_0)}}. \qquad (3.25)$$

As an illustration of this methodology, let us employ the steepest descent method to estimate the factorial $z!$ for a large integer value of z. We observe using a direct calculation (Feller, 1968) that

$$\ln(z!) = \ln 1 + \ln 2 + \cdots + \ln z$$

$$= \sum_{n=1}^{z} \ln n$$

$$\geqslant \int_{1}^{z} \ln(x)\, dx$$

$$= [x \ln x - x]_{1}^{z}$$

$$\geqslant z \ln(z) - z, \tag{3.26}$$

where we have employed integration by parts in obtaining the last line. Finally, this yields

$$z! \geqslant \left(\frac{z}{e}\right)^{z}, \tag{3.27}$$

a result often referred to as *Stirling's approximation*. However, we can obtain a more accurate result in applying the steepest descent method via Eq. (3.24) to

$$z! = \int_{0}^{\infty} x^{z} \exp(-x)\, dx, \tag{3.28}$$

from which we deduce that the steepest descent function $g(x)$ is

$$g(x) = x - z \ln x$$

$$g'(x) = 1 - \frac{z}{x}$$

$$g''(x) = \frac{z}{x^2}. \tag{3.29}$$

We identify that the minimum of $g(x)$ occurs when $x = z$. Making use of Eq. (3.25), we immediately obtain the result that

$$z! \approx \sqrt{2\pi z}\left(\frac{z}{e}\right)^{z}, \tag{3.30}$$

which is a substantially more accurate expression for the factorial function.

Before proceeding, we note a special case wherein $g(x)$ in (3.22) is a pure imaginary function and our methodology becomes known as the *method of stationary phase* (Bender and Orszag, 1999). We employ a similar quadratic expansion around a point x_0 where $g'(x_0)$ is zero and employ a suitably deformed contour that ultimately produces a Gaussian. This problem then becomes a special case of the steepest descent method. We will review contour integration in section 3.2. These integrals are related to the Stokes phenomenon and are the source of the phase shift in (2.116) for the Bessel function evaluated at large argument.

3.1.3 Special Problems in Geophysics; Elliptic Integrals

A commonly encountered problem in geophysics, and in analytical mechanics, emerges from integrals of the type (2.27) we saw applied to the pendulum problem where the potential had the form

$$V(x) = 1 - \cos(x) \tag{3.31}$$

and the associated quadrature giving the elapsed time t to swing from an angle x to the vertical as

$$t = \int_0^x \sqrt{\frac{m}{2[E - 1 + \cos(x')]}} \, dx'. \tag{3.32}$$

Integrals of this and related structures occur in many applications and have given rise to a class of problems known as *elliptic integrals*. For example, defining $\phi = x/2$ and $k^2 = 2/E$, the elapsed time satisfies

$$t = \sqrt{\frac{2E}{m}} \int_0^\phi \frac{d\varphi}{\sqrt{1 - k^2 \sin^2 \varphi}}. \tag{3.33}$$

Unfortunately, this rich topic is plagued by inconsistent notation and derivations that often require sophisticated applications of complex analysis and contour integration. Boas (2006) provides a brief summary of these functions and their uses. We synopsize some of the fundamental results here.

In the spirit of the previous equation, we define the *Legendre forms of the elliptic integrals* of the first and second kinds by

$$F(\phi, k) = \int_0^\phi \frac{d\varphi}{\sqrt{1 - k^2 \sin^2 \varphi}}, \quad 0 \leqslant k \leqslant 1,$$

$$E(\phi, k) = \int_0^\phi \sqrt{1 - k^2 \sin^2 \varphi} \, d\varphi, \quad 0 \leqslant k \leqslant 1. \tag{3.34}$$

There is a third kind of elliptic integral that is rarely encountered and will not be presented here. In addition to the Legendre forms presented above, there is also a *Jacobi form of the elliptic integrals* that emerges upon replacing $\sin \varphi$ by u and $\sin \phi$ by U. The forms presented above are often called "incomplete" because the angles involved are less than $\pi/2$ in the ϕ variable. The "complete" case corresponds to $\pi/2$, which, for the pendulum swing, corresponds to a vertical orientation.

Accordingly, we now define the complete (Legendre) forms of the elliptic integrals of the first and second kind, denoted by $K(k)$ and $E(k)$, respectively, namely,

$$K(k) = F\left(\frac{\pi}{2}, k\right) = \int_0^{\pi/2} \frac{d\varphi}{\sqrt{1 - k^2 \sin^2 \varphi}}$$

$$E(k) = E\left(\frac{\pi}{2}, k\right) = \int_0^{\pi/2} \sqrt{1 - k^2 \sin^2 \varphi} \, d\varphi. \tag{3.35}$$

Many textbooks provide details relating to the three types of elliptic integrals, albeit often in the context of complex analysis and contour integration. Jeffreys and Jeffreys (1999) and Whittaker and Watson (1979) are classic sources, as is Mathews and Walker (1970).

It should also be mentioned that there exists an inverse form commonly called *Jacobi elliptic functions*. To illustrate how this applies, consider as does Boas (2006) the following integral relation:

$$u = \int_0^x \frac{dt}{\sqrt{1 - t^2}} = \sin^{-1} x. \tag{3.36}$$

In that spirit, we now define

$$u = \int_0^x \frac{dt}{\sqrt{1 - t^2}\sqrt{1 - k^2 t^2}} = \mathrm{sn}^{-1} x. \tag{3.37}$$

By analogy, there is a form that resembles the cosine, and other quantities. The definitive treatment of these functions may be found as part of the "Bateman Project" (Bateman and Erdélyi, 1953). Their computation, especially using the *Landen transformation*, is detailed in the *NIST Handbook of Mathematical Functions* (Olver, 2010).

3.2   Complex Analysis and Elementary Contour Integration

This is a very rich topic and normally should be addressed in a regular lecture course. Here, we will simply review some of the fundamental definitions and results, and go on to employ the methods that emerge for calculating a broad class of integrals associated with mathematical physics in general, and geophysics in particular. An excellent source for review of these topics is Churchill (1960). In chapter 5, we will also employ these methods in application to integral transforms.

We consider a *complex variable* z as having the form $x + iy$, where both x and y are real and $i \equiv \sqrt{-1}$. The *complex conjugate* is denoted z^* and is equal to $x - iy$. The absolute value of z is denoted $|z|$ and is equal to $\sqrt{x^2 + y^2}$. Also, we say that $x = \Re z = \operatorname{Re} z$ and $y = \Im z = \operatorname{Im} z$. (Older books tend to use the $\Re$ and $\Im$ notation for the real and imaginary parts, while more current ones employ Re and Im.) A complex variable can also be expressed in polar form $z \equiv r \exp(i\theta)$ with $r = |z|$ and $x = r \cos(\theta)$ and $y = r \sin(\theta)$ since $\exp(i\theta) = \cos(\theta) + i \sin(\theta)$. A familiar result from complex analysis is *De Moivre's theorem* which is usually expressed as $\cos(n\theta) + i \sin(n\theta) = [\cos(\theta) + i \sin(\theta)]^n$ for integer-valued n.

Suppose a complex variable z can be mapped onto a complex variable w; we call w a function of the complex variable z and it is expressed

$$w = f(z). \tag{3.38}$$

A function is single valued if, for each value of z, there corresponds only one value of w. Otherwise, it is multiple valued. Commonly, we write

$$w = f(z) = u(x, y) + iv(x, y), \tag{3.39}$$

where u and v are real-valued functions of x and y. A very simple example of a single-valued function is given by

$$w = z^2 = (x + iy)^2 = x^2 - y^2 + i2xy, \tag{3.40}$$

so that

$$u(x, y) = x^2 - y^2$$
$$v(x, y) = 2xy. \tag{3.41}$$

It is evident here that $f(z)$ is single valued; when a function is multiple valued, we will show how it can be considered as a collection of single-valued functions.

The elementary concepts of limits and continuity encountered in the calculus of real variables can readily be extended to complex variables. Thus, $f(z)$ is said to have the *limit L* as z approaches z_0 if, given any $\epsilon > 0$, there exists a $\delta > 0$ such that $|f(z) - L| < \epsilon$ whenever $|z - z_0| < \delta$. Another way to express this is to say that $f(z)$ is continuous at z_0 if

$$\lim_{z \to z_0} f(z) = f(z_0). \tag{3.42}$$

Suppose now that $f(z)$ is continuous in some region of the z-plane. Then we say that the *derivative* of $f(z)$ denoted by $f'(z)$ is defined as

$$\frac{\mathrm{d}f(z)}{\mathrm{d}z} = f'(z) = \lim_{\Delta z \to 0} \frac{f(z + \Delta z) - f(z)}{\Delta z}, \tag{3.43}$$

provided that the limit exists *independent of the manner in which* $\Delta z \to 0$. If this limit exists for $z = z_0$, then we call $f(z)$ *analytic* or *regular* at z_0.

In many instances, we can define elementary functions of a complex variable by a natural extension of the corresponding functions of a real variable. For example, we may define $\exp(z) = 1 + z + z^2/2! + z^3/3! + \cdots$. However, not all functions of a complex variable z are single valued, just as is the case for real variables. For example, take

$$w(z) = z^{1/2}. \tag{3.44}$$

Suppose the original value point can be written $z = r\exp(i\theta)$. Then we observe that z will have two square roots $\sqrt{r}\exp(i\theta/2)$ as well as $\sqrt{r}\exp(i\theta/2 + i\pi)$. This follows since our original value of z could have been written $r\exp(i\theta)$ or $r\exp(i\theta + i2\pi)$. It is extremely important that we make our mapping single valued by selecting which of the two types of solution will be taken. Put another way, if we start at some point z and trace a closed curve in the z-plane, then we return to the position from which we started. The process of guaranteeing that we do not encounter the multivalued character of the function is to establish a *branch line* or *branch cut* from $z = 0$ to infinity, say along the negative real axis and agree never to cross it. The singular point $z = 0$ is called a *branch point*. The z-plane, when cut this way, is called a *Riemann sheet* of the function $w(z)$. It is useful to remember that $z = \infty$ is itself a single point in the complex plane, and we can often think of it as a branch point. Two branch points can be joined together by a branch cut or line; the Riemann sheet that we have just constructed is a case in point.

The *Cauchy–Riemann equations* emerge from the (necessary) condition that $w = f(z) = u(x, y) + iv(x, y)$ be analytic or regular in a region $\mathcal{R}$. In particular, the value of the derivative should be independent of the direction chosen. Thus, by allowing Δz to be in the x-direction and then in the y-direction, it is easy to show that

$$\frac{\partial u}{\partial x} = +\frac{\partial v}{\partial y}$$
$$\frac{\partial u}{\partial y} = -\frac{\partial v}{\partial x}, \tag{3.45}$$

first by taking Δz in the x-direction and then by taking Δz in the y-direction. By taking the second derivative of both u and v with respect to x and to y, we find upon differentiating that

$$\frac{\partial^2 u}{\partial x^2} + \frac{\partial^2 u}{\partial y^2} = 0$$
$$\frac{\partial^2 v}{\partial x^2} + \frac{\partial^2 v}{\partial y^2} = 0. \tag{3.46}$$

Thus, we see that both the real and imaginary parts of *any* analytic function satisfy Laplace's equation in two dimensions. Functions that satisfy Laplace's equations are called *harmonic functions*.

Because analytic functions have derivatives that are independent of the direction in which the point in our definition is approached, analytic functions also satisfy some remarkable properties under integration. Suppose we are integrating $f(z)$ along some path C from point z_1 to z_2, where $z_j = x_j + iy_j$, for $j = 1, 2$. The associated integral

$$I_C = \int_{z_1}^{z_2} f(z) \, dz \tag{3.47}$$

is a line integral that in principle depends on the path C followed from z_1 to z_2. However, the integral will be the same for *any* paths if z is regular in the region bounded by the paths. This is equivalent to what is known as *Cauchy's theorem*:

$$\oint_C f(z) \, dz = 0, \tag{3.48}$$

for *any* closed path lying within the region in which $f(z)$ is analytic or regular. (The proof of this result is very similar to that employed for establishing Green's theorem in the context of line integrals, and can be found in most complex analysis textbooks.) Conversely, if this line integral vanishes for *every* closed path within a region where $f(z)$ is continuous and single valued, then $f(z)$ is regular in that region.

An additional result that emerges here is that, if $f(z)$ is regular in a region, then its derivatives of all orders exist and are regular there. Analytic behavior brings with it some strong benefits, one of which being *Cauchy's integral formula*

$$f(z) = \frac{1}{2\pi i} \oint_C \frac{f(\zeta)\, d\zeta}{\zeta - z}. \tag{3.49}$$

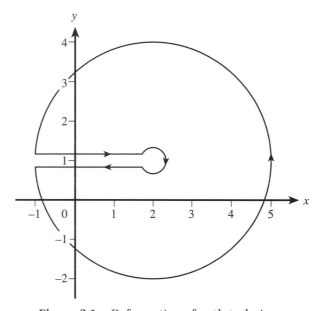

Figure 3.1. Deformation of path technique.

By convention, we assume that the direction of integration is counterclockwise.

Suppose that $f(z)$ is regular everywhere inside the region enclosing the unusual contour shown in Figure 3.1. However, let us imagine that we select a point $z = 2 + i$ in the complex plane, so the integrand in (3.49) in the complex variable ζ is not regular at that point. Accordingly, the closed path of integration shown there contains a region that is regular, thereby making the integral shown vanish. We will now deconstruct the contour into four parts: Note that the two parallel line segments are assumed to be infinitesimally close together. Similarly, the two almost-complete circles contain a gap of the same size that is also infinitesimal in extent. We observe that the integration along the two line segments, in the limit of an infinitesimal separation, exactly cancels. The large-radius circle is representative of any finite-size closed curve (apart from the infinitesimal gap). The inner circle in the ζ plane, with assumed infinitesimal radius enclosing the point z, therefore contribute as much—albeit with a negative sign since the direction of integration is clockwise—as the outer circle denoting *any* closed contour. This, accordingly, makes the value of the contour integral associated with the large circle the same as that due to the infinitesimal enclosed circle. We will now validate the Cauchy integral formula (3.49) in the vicinity of the singularity.

We will express ζ as $z + \epsilon \exp(i\theta)$, thereby making $d\zeta = i\epsilon \exp(i\theta) d\theta$. Since $f(\zeta)$ is regular, it can be expressed as a Taylor series $f(z) + (\zeta - z)f'(z)$, and higher order terms. The right-hand side of (3.49) then becomes

$$\frac{1}{2\pi i} \oint_c \frac{f(\zeta) \, d\zeta}{\zeta - z}$$

$$= \lim_{\epsilon \to 0} \frac{1}{2\pi i} \oint_c \frac{f(z) + \epsilon \exp(i\theta) f'(z)}{\epsilon \exp(i\theta)} i\epsilon \exp(i\theta) \, d\theta$$

$$= \frac{1}{2\pi i} f(z) 2\pi i = f(z), \tag{3.50}$$

as desired. Much of complex analysis and its application to the evaluation of integrals is associated with the construction of appropriate contours that yield the contribution desired in terms of contributions emerging from isolated singularities.

Cauchy's formula (3.49) may be differentiated as often as we like to obtain the formula

$$f^{(n)}(z) = \frac{n!}{2\pi i} \oint_C \frac{f(\zeta) \, d\zeta}{(\zeta - z)^{n+1}}, \tag{3.51}$$

for $n = 1, 2, \ldots$ Consequently, a power series expansion or Taylor's series can be constructed around any point z_0 in a region where $f(z)$ is regular, that is,

$$f(z) = f(z_0) + \frac{1}{1!} f^{(1)}(z_0)(z - z_0) + \frac{1}{2!} f^{(2)}(z_0)(z - z_0)^2 + \cdots . \tag{3.52}$$

We refer to the region in the z-plane in which the series converges as a circle, with the *circle of convergence* extending to the nearest point where $f(z)$ ceases to be analytic. Moreover, if the power series converges within some circle, then $f(z)$ is regular there.

At this juncture, we need to introduce some additional terminology. A *singular point* of a function $f(z)$ is a value of z at which $f(z)$ is not analytic. If $f(z)$ is analytic everywhere in some region except at an interior point, say $z = a$, we call $z = a$ an *isolated singularity* of $f(z)$. For example, if $f(z) = 1/(z - 2)^3$, then $z = 2$ is an isolated singularity. Suppose

$$f(z) = \frac{\psi(z)}{(z - a)^n}, \tag{3.53}$$

where $\psi(z)$ is analytic everywhere in a region including $z = a$ but where $\psi(a) \neq 0$ and n is a positive integer. We say that $f(z)$ has an isolated singularity at $z = a$, which is called a *pole of order n*. If $n = 1$, we call it a *simple pole*; if $n = 2$, we call it a *double pole*; and so on. As an illustration, consider

$$f(z) = \frac{z^2}{(z - 2)^3(z + 2)}. \tag{3.54}$$

We say that $f(z)$ has two singularities, a pole of order 3 or a triple pole at $z = 2$ and a simple pole at $z = -2$.

Singularities can appear in ways other than poles. We already observed that $f(z) = \sqrt{z}$ has a branch point at $z = 0$. Another case emerges when we consider $f(z) = \sin(z)/z$: While it has a singularity at $z = 0$, its limit exists there and is finite, rendering the origin a *removable singularity*. In addition to power-series expansions, the so-called *Laurent expansion* is of great value in dealing with isolated singularities. If $f(z)$ is regular in an annulus situated between two concentric circles centered at z_0, then $f(z)$ may be represented within this region by a Laurent expansion

$$f(z) = \sum_{n=-\infty}^{\infty} a_n(z - z_0)^n, \tag{3.55}$$

where the coefficients a_n satisfy for $n \geqslant 0$

$$a_n = \frac{1}{2\pi i} \oint_C \frac{f(z)\,dz}{(z - z_0)^{n+1}}, \tag{3.56}$$

where C is *any* closed path encircling z_0 counterclockwise within the annular region. Note also that

$$a_{-1} = \frac{1}{2\pi i} \oint_C f(z)\,dz. \tag{3.57}$$

A Laurent expansion can be employed to establish that z_0 be a pole of order m by multiplying the expansion by $(z - z_0)^m$. If z_0 is a pole of order m, a convergent power series for $(z - z_0)^m f(z)$ emerges since that product will be analytic near z_0 and

$$\lim_{z \to z_0} (z - z_0)^m f(z) \neq 0; \qquad (3.58)$$

if z_0 is an *essential singularity*, then $(z - z_0)^m f(z)$ is unbounded near z_0 for all positive integer values m. If z_0 is an isolated singularity, the coefficient a_{-1} shown explicitly in (3.57) is called the *residue* of $f(z)$ at z_0. The presence of residues is of profound importance to the calculation of integrals.

The *residue theorem* facilitates the evaluation of an integral of a function $f(z)$ along a closed path C such that $f(z)$ is regular in the region bounded by C except for a finite number of poles and (isolated) essential singularities in the interior of C. By Cauchy's theorem, as illustrated earlier via Figure 3.1, the path or *contour* may be deformed without crossing any singularities until it is reduced to little circles surrounding each singular point. Suppose $f(z)$ has simple pole at a set of points z_i inside a closed contour. Then the residue at z_i can be calculated according to

$$\text{residue}_i = \lim_{z \to z_i} (z - z_i) f(z). \qquad (3.59)$$

The integral around each such little circle is then given by (3.57), yielding the theorem of residues:

$$\oint_C f(z)\,dz = 2\pi i \sum \text{residues}_i \qquad (3.60)$$

where the sum is over all of the poles z_i and essential singularities inside C. If a pole happens to lie on a contour, we have to investigate whether this is an artifact of some approximation and, if so, on which side of the contour. We present next an example, and recommend textbooks such as Churchill (1960) for other illustrative examples.

Consider now the integral

$$\mathcal{I} = \frac{1}{2} \int_0^{2\pi} \frac{d\theta}{a + b\cos\theta}, \qquad a > b > 0. \qquad (3.61)$$

Integrals of trigonometric type are often evaluated by making use of the transformation

$$z = \exp(i\theta), \quad dz = i\exp(i\theta)\,d\theta. \qquad (3.62)$$

and employ

$$\cos\theta = \frac{\exp(i\theta) + \exp(-i\theta)}{2} = \frac{1}{2}\left(z + \frac{1}{z}\right). \qquad (3.63)$$

Then, our integral (3.61) becomes

$$\mathcal{I} = \int_C \frac{dz}{iz}\frac{1}{2a + b(z + 1/z)} = \frac{1}{i}\oint_C \frac{dz}{bz^2 + 2az + b}, \qquad (3.64)$$

where the contour is the (counterclockwise) unit circle in z.

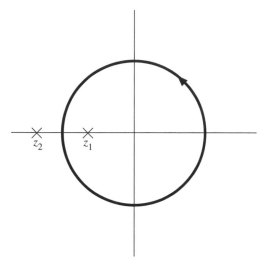

Figure 3.2.   Contour employed in z-plane.

We observe that the quadratic in the integrand's denominator has two solutions, namely,

$$z_{1,2} = -\frac{a}{b} \pm \sqrt{\frac{a^2}{b^2} - 1}, \qquad (3.65)$$

where the positive case corresponds to subscript 1 and the negative case to subscript 2. Since only z_1 is enclosed by the contour, as shown in Figure 3.2, we need only evaluate its residue. Therefore, we observe that

$$\mathcal{I} = \frac{1}{i}2\pi i \times \text{residue at } z_1 = 2\pi\frac{1}{2bz_1 + 2a} = \frac{\pi}{\sqrt{a^2 - b^2}}. \qquad (3.66)$$

With this example, we conclude our brief review of complex analysis and contour integration. We observe that these techniques provide a powerful addition to the approximate methods discussed earlier in this chapter. We now proceed to discuss integral transforms, which, as it happens, will make good use of these methods!

3.3 Fourier Transforms and Analysis Methods

In this section, we introduce fundamental ideas relating to Fourier series and their transforms, as well as Fourier transforms of square-integrable functions. We advance to Fourier transform pairs and multidimensional transforms. We introduce the sampling theorem and methods to combat the potential for aliasing of transforms. We conclude this section by introducing approximation methods, the discrete Fourier transform, and its revolutionary adaptation, the fast Fourier transform.

3.3.1 Fourier Series, Transforms, and Convolutions

Let us consider a complex function of time, $x(t)$, defined, for convenience, over $[-T/2, T/2]$, and we will assume that

$$\int_{-T/2}^{T/2} |x(t)|^2 \, dt = \int_{-T/2}^{T/2} x(t)x^*(t) \, dt < \infty.$$

For convenience, we will from time to time refer to this as our "signal." Our presentation here is very brief, and there are significant details that we are omitting. However, these can generally be found in an advanced undergraduate or graduate textbook. We have assumed that $x(t)$ is complex owing to the greater generality that it provides. It is useful to think of $x(t)$ as though it were a measured voltage in an experiment. Accordingly, its square is proportional to power and the corresponding integral over all time, which is presumed finite, is the energy present in the signal. Since $x(t)$ is presumed to be known, it represents a *deterministic signal*, in contrast with a *stochastic signal* where the repetition of the "experiment" employed to measure it would recover a different signal that preserves certain statistical properties. In stochastic applications, the signal is often presumed to be *stationary*, that is, its statistical properties do not change over time. Inasmuch as the signal we are considering here is deterministic and represents a finite amount of energy, it is *nonstationary*, a term that is used in some applications to describe some stochastic processes.

From the theory of Fourier series (Strang, 1986; Walker, 1988; Mathews and Walker, 1970), we can write

$$x(t) = \frac{1}{T} \sum_{n=-\infty}^{\infty} a_n \exp\left(-2\pi i \frac{nt}{T}\right). \tag{3.67}$$

Note that we have, without loss of generality, chosen to center the range of time at the origin, and we employed complex exponentials, instead of sines and cosines, that are a complex function of time, since both sets of functions are equivalent and complete basis sets. The a_n coefficients can be obtained by minimizing the quantity

$$U \equiv \int_{-T/2}^{T/2} \left| x(t) - \frac{1}{T} \sum_{n=-\infty}^{\infty} a_n \exp\left(-2\pi i \frac{nt}{T}\right) \right|^2 dt, \qquad (3.68)$$

from which we obtain that, for $n = -\infty, \ldots, \infty$,

$$a_n = \int_{-T/2}^{T/2} x(t) \exp\left(+2\pi i \frac{nt}{T}\right) dt. \qquad (3.69)$$

We ultimately wish to extend the range $T \to \infty$, so we introduce a new frequency variable v in discrete increments $v_n \equiv n/T$ and observe that its increment with respect to n or $\Delta v_n = \Delta n/T = 1/T$. Accordingly, it is appropriate to associate a_n with the complex amplitude at v_n, which we will denote as $f(v_n)$. Hence,

$$a_n \to f(n/T) = f(v_n) = \lim_{T \to \infty} \int_{-T/2}^{T/2} x(t) \exp(2\pi i v_n t)\, dt, \quad (3.70)$$

or, since we no longer need to reference n,

$$f(v) = \int_{-\infty}^{\infty} x(t) \exp(2\pi i v t)\, dt. \qquad (3.71)$$

Similarly, we observe that we can write

$$x(t) = \sum_{n=-\infty}^{\infty} f(v_n) \exp(-2\pi i v_n t) \Delta v_n$$

$$\to \int_{-\infty}^{\infty} f(v) \exp(-2\pi i v t)\, dv. \qquad (3.72)$$

Our derivation could have employed other normalizations, but would have given the same outcome, namely, the Fourier transform pair

$$x(t) = \int_{-\infty}^{\infty} f(v) \exp(-2\pi i v t)\, dv$$

$$f(v) = \int_{-\infty}^{\infty} x(t) \exp(+2\pi i v t)\, dt. \qquad (3.73)$$

It is quite common to use the angular frequency $\omega \equiv 2\pi v$ instead, whereupon we obtain

$$x(t) = \frac{1}{2\pi} \int_{-\infty}^{\infty} f(\omega) \exp(-i\omega t)\, d\omega$$

$$f(\omega) = \int_{-\infty}^{\infty} x(t) \exp(+i\omega t)\, dt, \qquad (3.74)$$

where the argument of the Fourier transform f has been redefined, where we have changed the argument of the frequency-related function to the angular frequency, that is, $a(\omega)$, in keeping with these definitions. The placement of the normalizing 2π term is arbitrary; so long as the product of the terms multiplying the respective integrals is $1/2\pi$, the definitions are self-consistent. We will employ both notations throughout this chapter, selecting the one that simplifies the calculation.

3.3.2 Illustrative Examples of Fourier Transform Pairs

Four special examples of Fourier transform pairs merit particular attention.

1. In the first, we will assume that $x(t)$ is the δ function $\delta(t)$. We immediately observe that $f(v) = 1$. Similarly, if $f(v) = \delta(v)$, then $x(t) = 1$. This situation shows that a monochromatic signal will preserve the signal indefinitely in time. (Note that this example is no longer square integrable, a reflection of the singularity associated with the δ function.) An important corollary of this result is that

$$\delta(\tau) = \int_{\infty}^{\infty} \exp(-2\pi i v \tau)\, dv. \qquad (3.75)$$

One other feature of the δ function should be noted here. Since

$$\int_{-\infty}^{\infty} \delta(t) g(t)\, dt = g(0), \qquad (3.76)$$

we need to address the issue that emerges when the argument of the δ function is itself a function, say $h(t)$. Let us consider

$$\int_{-\infty}^{\infty} \delta[h(t)] g(t)\, dt = \int_{-\infty}^{\infty} \frac{\delta[h(t)] g(t) |dh(t)/dt|}{|dh(t)/dt|}\, dt$$

$$= \int \delta[h] g[t(h)] \frac{d|h(t)|}{|h'[t(h)]|}, \qquad (3.77)$$

where we have assumed, at all times t_i where $h(t_i) = 0$, that the derivative of $h(t)$ is nonzero so that we can define locally $t(h)$. Proceeding accordingly, we observe that

$$\int_{-\infty}^{\infty} \delta[h(t)]g(t)\,dt = \sum_i \frac{g(t_i)}{|h'(t_i)|}, \qquad (3.78)$$

where the summation is performed over all times t_i where $h(t)$ vanishes.

2. In the second example, we will assume that $x(t)$ has the form of a Gaussian $\exp(-\beta t^2)$, thereby giving

$$f(\nu) = \int_{-\infty}^{\infty} \exp(-\beta t^2)\exp(2\pi i\nu t)\,dt. \qquad (3.79)$$

To evaluate this integral, we first complete squares inside the exponential, namely,

$$f(\nu) = \int_{-\infty}^{\infty} \exp\left\{-\beta\left[\left(t - \frac{\pi i\nu}{\beta}\right)^2 + \frac{\pi^2\nu^2}{\beta^2}\right]\right\}\,dt$$

$$= \exp\left(-\frac{\pi^2\nu^2}{\beta}\right)\int_{-\infty}^{\infty}\exp\left[-\beta\left(t - \frac{\pi i\nu}{\beta}\right)^2\right]\,dt. \quad (3.80)$$

We now deform the contour in the complex t-plane accordingly since the integrand contains no singularities in the vicinity of the real t-axis.

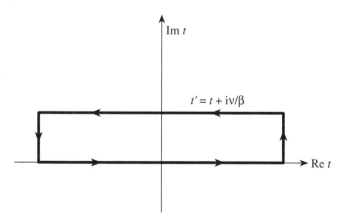

Figure 3.3. Contour employed in the t-plane for Fourier transform of a Gaussian.

We consider the transformation $t' = t + i\pi\nu/\beta$ and obtain

$$f(\nu) = \exp\left(-\frac{\pi^2\nu^2}{\beta}\right)\int_{-\infty+i\pi\nu/\beta}^{\infty+i\pi\nu/\beta}\exp[-\beta t'^2]\,dt'. \qquad (3.81)$$

Figure 3.3 illustrates this: Normally, we would perform the integral on the real t-axis but observe that it is more natural to perform it on the displaced t'-axis. The end members of the contour are infinitely far removed from the origin and, since they contain $\exp(-\beta t^2)$ terms, are infinitesimal and do not contribute. Therefore, we conclude that (3.81) is equal to the integral performed on the undeformed path. Finally, we observe that

$$f(v) = \sqrt{\frac{\pi}{\beta}} \exp\left(-\frac{\pi^2 v^2}{\beta}\right). \tag{3.82}$$

Importantly, the original Gaussian function of time has been transformed into a Gaussian function of frequency. Moreover, we note that the "width" of the original Gaussian (3.79) is inversely proportional to the width of the Gaussian (3.82), with the proportionality constant being π. The relationship between the width of a distribution and its Fourier inverse is known as the *uncertainty principle* (Bracewell, 2000) and can be rigorously established as a theorem. It provides the basis for the quantum mechanical principle by the same name. The uncertainty principle in physics is often misunderstood and misconstrued as being an artifact of physical boundaries, whereas in fact it is an outcome of Fourier theory.

3. We will define a *boxcar* function by

$$x(t) = \begin{cases} 1 & \text{if } |t| < T/2, \\ 0 & \text{if } |t| > T/2. \end{cases} \tag{3.83}$$

Then, we obtain

$$f(v) = \frac{\sin(\pi v T)}{\pi v} = T \operatorname{sinc}(\pi v T), \tag{3.84}$$

where we have now introduced the *sinc* function

$$\operatorname{sinc}(x) \equiv \frac{\sin(x)}{x}. \tag{3.85}$$

The sinc function is commonly employed in a variety of signal-processing applications and is generally the outcome of a source of data being "windowed," that is, visible over some range of time but otherwise not visible. Another aspect of Fourier transforms evident from this result is that a sharp edge in the time domain, as would emerge with a boxcar filter, results in energy being distributed over all frequencies.

4. The output emergent from an RC circuit shows an exponential decay with a decay time of $\tau = RC$, thus

$$x(t) = \begin{cases} 0 & \text{if } t < 0, \\ \exp(-t/\tau) & \text{if } t \geqslant 0. \end{cases} \qquad (3.86)$$

Accordingly, its Fourier transform is

$$f(v) = \frac{\tau}{1 - 2\pi i v \tau} = \frac{i}{2\pi(v + i/2\pi\tau)}, \qquad (3.87)$$

which reveals that $f(v)$ has a simple pole in its lower half-plane. We will see shortly that this property has important consequences, for example, in the context of the Laplace transform.

5. As a further illustration of the utility of Fourier transform pairs, consider now the initial value problem presented by the heat or diffusion equation in one dimension:

$$\frac{\partial T(x,t)}{\partial t} = D \frac{\partial^2 T(x,t)}{\partial x^2}, \qquad (3.88)$$

for $-\infty < x < \infty$ and initial conditions

$$T(x,0) = T_0(x). \qquad (3.89)$$

Let us introduce a transform quantity $F(k,t)$ and the associated Fourier transforms

$$F(k,t) = \frac{1}{2\pi} \int_{-\infty}^{\infty} \exp(-ikx) T(x,t) \, dx$$

$$T(x,t) = \int_{-\infty}^{\infty} \exp(ikx) F(k,t) \, dk. \qquad (3.90)$$

Employing the initial condition, we observe that we can write

$$F(k,0) = \frac{1}{2\pi} \int_{-\infty}^{\infty} \exp(-iky) T_0(y) \, dy. \qquad (3.91)$$

We apply the heat equation (3.88) to the second Fourier equation in (3.90) and observe that

$$\frac{\partial F(k,t)}{\partial t} = -Dk^2 F(k,t), \qquad (3.92)$$

and we immediately have the solution

$$F(k,t) = \exp(-Dk^2 t)F(k,0). \qquad (3.93)$$

Hence, we can write

$$T(x,t)$$

$$= \int_{-\infty}^{\infty} \exp(ikx)F(k,0)\exp(-Dk^2 t)\,dk$$

$$= \frac{1}{2\pi} \int_{-\infty}^{\infty} \int_{-\infty}^{\infty} \exp[ik(x-y)]T_0(y)\exp(-Dk^2 t)\,dk\,dy.$$

$$(3.94)$$

We now exploit the result obtained in our second example in performing the k integral, which is a Fourier transform of a Gaussian, and obtain

$$T(x,t) = \frac{1}{\sqrt{4\pi Dt}} \int_{-\infty}^{\infty} dy\,T_0(y)\exp\left[-\frac{(x-y)^2}{4Dt}\right]. \quad (3.95)$$

We now identify the kernel of this integral, namely, the terms in the integrand apart from the initial condition $T_0(y)$ according to

$$G(x,t;y) = \frac{1}{\sqrt{4\pi Dt}}\exp\left[-\frac{(x-y)^2}{4Dt}\right], \qquad (3.96)$$

as the Green's function for the heat equation since $G(x,t)$ can readily be shown to satisfy the heat equation (3.88) and observe that

$$\lim_{t\to 0} G(x,t;y) = \delta(x-y), \qquad (3.97)$$

since in that limit $G(x,t;y)$ will vanish when $x \neq y$ and, upon integration over y, has unit value. A useful interpretation of this result is that G provides the response from a "unit" impulse of heat at y as observed at x at a later time t.

An excellent resource and compilation of transform pairs can be found in Bracewell (2000).

3.3.3 Multidimensional and Other Fourier Transform Pairs

We wish to address here some issues that emerge in the construction of multidimensional Fourier transforms. Our focus for the moment will be two dimensional, but we will return to the full three-dimensional problem in chapter 4. Suppose we have a function $f(x, y)$ whose Fourier transform $F(u, v)$ we wish to determine. The transform pair often employed, especially in radio astronomical applications (Bracewell, 2000), extends our formulation (3.73) to the pair

$$f(x, y) = \int_{-\infty}^{\infty} \int_{-\infty}^{\infty} F(u, v) \exp[-2\pi i(ux + vy)] \, du \, dv$$

$$F(u, v) = \int_{-\infty}^{\infty} \int_{-\infty}^{\infty} f(x, y) \exp[+2\pi i(ux + vy)] \, dx \, dy,$$

$$(3.98)$$

and astronomers talk about coverage in the u-v plane. Another formulation emerges from electromagnetic theory employing vector positions $\boldsymbol{x}$ and wavenumber $\boldsymbol{k}$, thus

$$f(\boldsymbol{x}) = \frac{1}{2\pi} \int_{-\infty}^{\infty} \int_{-\infty}^{\infty} F(\boldsymbol{k}) \exp[+i\boldsymbol{k} \cdot \boldsymbol{x}] \, d^2 k$$

$$F(\boldsymbol{k}) = \frac{1}{2\pi} \int_{-\infty}^{\infty} \int_{-\infty}^{\infty} f(\boldsymbol{x}) \exp[-i\boldsymbol{k} \cdot \boldsymbol{x}] \, d^2 x, \qquad (3.99)$$

where we have symmetrized the normalizations in the two expressions. In many applications, it is advantageous to employ polar coordinates, with $f(x, y) \to f(r, \theta)$. The latter, in turn, can be adapted using Fourier series, namely,

$$f(r, \theta) = \sum_{-\infty}^{\infty} f_n(r) \exp(in\theta), \qquad (3.100)$$

where the "azimuthal" modes $f_n(r)$ can be calculated by

$$f_n(r) = \frac{1}{2\pi} \int_0^{2\pi} f(r, \theta) \exp(-in\theta) \, d\theta. \qquad (3.101)$$

In two dimensions, the kernel $\exp(-i\boldsymbol{k} \cdot \boldsymbol{x})$ can be expressed as a series that arises from the generating function for Bessel functions, which we presented in chapter 2, namely,

$$\exp\left[\frac{1}{2}z\left(t - \frac{1}{t}\right)\right] = \sum_{m=-\infty}^{\infty} t^m J_m(z), \qquad (3.102)$$

which, upon substituting $\exp(i\theta)$ for t, yields

$$\exp[iz\sin(\theta)] = \sum_{m=-\infty}^{\infty} \exp(im\theta) J_m(z). \qquad (3.103)$$

A particularly important case emerges when $f(\boldsymbol{x})$ is circularly symmetric, that is, depends only upon the radius r but not on the angle θ, that is, $f(r, \theta) = f_0(r)$. Then, $F(\boldsymbol{k})$ becomes simply a function $F(k)$ of the length k of the $\boldsymbol{k}$-vector and

$$F(k) = \frac{1}{2\pi} \int_0^{2\pi} \int_0^\infty f_0(r) \exp(-ikr \cos \theta) r \, dr \, d\theta. \quad (3.104)$$

Note, if $f(\boldsymbol{x})$ were not circularly symmetric, then $F(\boldsymbol{k})$ would have a Fourier series representation analogous to (3.100). Using (3.103), the latter reduces to

$$F(k) = F_0(k) = \int_0^\infty f_0(r) J_0(kr) r \, dr. \quad (3.105)$$

In analogous fashion, we can derive

$$f_0(r) = \int_0^\infty F_0(k) J_0(kr) k \, dk. \quad (3.106)$$

The latter pair of equations are referred to as the *Fourier–Bessel transform* as well as the *Hankel transform*. Equivalent expressions can be derived for all azimuthal modes $f_n(r)$ and $F_n(k)$. We will return to polar coordinate geometry issues later in this chapter.

Fourier transform methods in one dimension emerge in other contexts depending on the nature of the poles present in the problem. In the case of a simple pole being present exactly on the real axis and where there is no clear choice available regarding whether to deform the contour above the pole or below it, it is customary to take the average of the two results, that is, take 1/2 of the residue associated with that pole plus the contribution to the integral emerging from the line integral on both sides of the pole. This is called the *principal value*. When this occurs in the context of performing a Fourier transform, the product of the transform also has a simple pole present on its axis and a similar procedure is needed to invert the transform. This gives rise to the so-called *Hilbert transform* pair (Mathews and Walker, 1970).

In other situations, the Fourier transform of a function is not available because it is not square integrable. In situations like that, we typically set the function to zero before the origin, and modulate it thereafter with an exponential decay envelope to render it square integrable, that is, in place of a function $f(x)$, we now consider $f(x)H(x)\exp(-cx)$, where $H(x)$ is the Heaviside function. In so doing, we have created a *Laplace transform* (Mathews and Walker, 1970). This form of transform is not encountered frequently in physics-based applications, but is often helpful in engineering in solving linear differential equations. The quantity we have constructed has the Fourier transform

$$g(y) = \int_{-\infty}^{\infty} f(x)\exp(-cx)H(x)\exp(-ixy)\,dx$$

$$= \int_{0}^{\infty} f(x)\exp(-cx)\exp(-ixy)\,dx. \qquad (3.107)$$

Given our expression (3.73), the inverse transform is formally easy to express, but this outcome is best accomplished using the so-called *Bromwich integral* (Mathews and Walker, 1970; Arfken and Weber, 2005). Since this methodology is rarely used in geophysics, we will not provide further details here.

We return now to the Fourier transform properties of a deterministic signal $x(t)$. It is evident that $f(v)$ represents the amplitude and (complex) phase of the frequency v in its contribution to the signal, so it is natural to think of it as being a measure of the "voltage" at that frequency, just as we took $x(t)$ to be the voltage at that time. Accordingly, using (3.73), it is natural to define the *power spectrum* $S(v)$ by

$$S(v) \equiv |f(v)|^2 = f(v)f^*(v). \qquad (3.108)$$

In analogy to the Fourier transform pairs above, we wish to construct the inverse transform for $S(v)$, which we will call the *autocorrelation function* $\rho(t)$, so that we have the pair

$$\rho(t) = \int_{-\infty}^{\infty} S(v)\exp(-2\pi ivt)\,dv$$

$$S(v) = \int_{-\infty}^{\infty} \rho(t)\exp(+2\pi ivt)\,dt. \qquad (3.109)$$

We wish to identify a direct relationship between $\rho(t)$ and $x(t)$. Using our definition for $f(v)$, we find that

$$\rho(t) = \int_{\infty}^{\infty}\int_{\infty}^{\infty}\int_{\infty}^{\infty} x(t')x^*(t'')\exp[2\pi iv(-t+t'-t'')]\,dt''\,dt'\,dv.$$
$$(3.110)$$

Invoking (3.75), we observe that this becomes

$$\rho(t) = \int_{-\infty}^{\infty} \int_{-\infty}^{\infty} x(t')x^*(t'')\delta(t - t' + t'')\,\mathrm{d}t'\,\mathrm{d}t''$$

$$= \int_{-\infty}^{\infty} x(t')x^*(t' - t)\,\mathrm{d}t'$$

$$= \int_{-\infty}^{\infty} x(t'' + t)x^*(t'')\,\mathrm{d}t'', \tag{3.111}$$

where we integrated first over t'' and then replaced t' by $t'' + t$.

The following schematic illustrates through the use of arrows important elements of the transformations we have discussed. Double arrows imply that one can proceed in either direction. So, we can reversibly go from $x(t)$ to $f(v)$ and back, just as we can reversibly go from $\rho(t)$ to $S(v)$ and back. However, the single arrows (pointing down) indicate that passage from one variable to the other can proceed only in one direction. We can proceed from $x(t)$ to $\rho(t)$, but cannot return due to information lost in calculating the autocorrelation function integral. Similarly, we can proceed from $f(v)$ to $S(v)$, but cannot return due to phase information lost in calculating $f(v)f^*(v)$.

$$x(t) \iff f(v)$$
$$\Downarrow \qquad \Downarrow$$
$$\rho(t) \iff S(v)$$

One final feature of our transform pairs is that we can obtain $\rho(t)$ without directly performing an autocorrelation. In particular, we proceed via a Fourier transform from $x(t)$ to $f(v)$, then calculate $S(v) = f(v)f^*(v)$, and then perform a Fourier inverse transform to recover $\rho(t)$. We will later observe, in practical applications involving data, that the availability of the fast Fourier transform makes this circuitous route much more efficient.

We observe from (3.111) that

$$\rho(0) = \int_{-\infty}^{\infty} x(t)x^*(t)\,\mathrm{d}t \tag{3.112}$$

but from (3.109) that

$$\rho(0) = \int_{-\infty}^{\infty} f(v)f^*(v)\,\mathrm{d}v. \tag{3.113}$$

The equivalence of these two expressions has a beautiful physical interpretation.

Parseval's theorem emerges by equating the latter two equations. Intuitively, each integral describes the energy associated with two different measures of the "energy" resident in the signal, the first being the time integral over the power (energy per time units) and the second being the frequency integral over the power ("spectral density" or energy per frequency units).

Before departing from this section, this is an appropriate time to introduce the topic of *convolutions*. Analogous to (3.75), let us define a convolution $y(t)$ of $x(t)$ with $h(t)$, namely,

$$y(t) = \int_{-\infty}^{\infty} x(t + t')h^*(t')\, dt'; \tag{3.114}$$

usually, we are concerned with real-valued quantities but we are preserving here the possibility of complex-valued quantities for the sake of completeness. Suppose $x(t)$ describes some physical process and $h(t')$ describes how that process is "smoothed" so that $y(t'')$ describes the observable outcome. For convenience, we will assume that their respective Fourier transforms are $X(\nu)$, $\mathcal{H}(\nu)$, and $Y(\nu)$. We take the inverse transform to obtain

$$Y(\nu) = \int_{-\infty}^{\infty} y(t) \exp(2\pi i \nu t)\, dt$$

$$= \iint dt\, dt'\, \exp(2\pi i \nu t)x(t + t')h^*(t')$$

$$= \iiiint dt\, dt'\, d\nu'\, d\nu''\, \exp(2\pi i \nu t)X(\nu')$$

$$\cdot \exp[-2\pi i \nu'(t + t')]\mathcal{H}^*(\nu'') \exp(2\pi i \nu'' t'), \tag{3.115}$$

where the integrals are from $-\infty$ to ∞. After separating the exponentials into four terms, we perform the t' integration, which recovers $\delta(\nu'' - \nu')$, and integrate over ν'' to obtain

$$Y(\nu) = \iint dt\, d\nu'\, \exp[2\pi i(\nu - \nu')t]X(\nu')\mathcal{H}^*(\nu')$$

$$= \int d\nu'\, \delta(\nu - \nu')X(\nu')\mathcal{H}^*(\nu')$$

$$= X(\nu)\mathcal{H}^*(\nu), \tag{3.116}$$

where all of the integrals have collapsed, owing to the repeated emergence of δ functions. This result, together with (3.114), is known as the *convolution theorem* (Mathews and Walker, 1970) and is referred to in some sources as the *faltung*, the German word for "convolution."

Many circumstances in nature give rise to convolution. The transmission of a signal through an RC filter convolves the original signal with

$$h(t) = \begin{cases} 0 & \text{if } t < 0, \\ \exp(-t/\tau) & \text{if } t \geqslant 0, \end{cases} \tag{3.117}$$

where $h(t)$ is called a *transfer function*. Importantly, for that case, we have already derived its Fourier transform,

$$\mathcal{H}(v) = \frac{\tau}{1 - 2\pi i v \tau}. \tag{3.118}$$

Many geophysical environments yield delays in propagation owing, for example, to inhomogeneities yielding "multipath" propagation. In the field of seismology, *coda* (Aki and Richards, 2002; Shearer, 2009) results in the received signal $y(t)$ being an amalgam of many different paths, each presenting different times of propagation. In atmospheric physics, inhomogeneities have the same effect and give rise to *scintillation*, the "twinkling" of stars. Variability in interplanetary and interstellar medium (plasma) densities results in dispersive propagation and the equivalent of scintillation in radio frequencies received from both solar system and extra-solar (including galactic) sources.

A natural question that emerges is whether one can take the measured signal $y(t)$ and, knowing the transfer function, identify the signal $x(t)$ at source. In principle, this is possible by using (3.116) to "deconvolve" the received signal, that is, apply

$$X(v) = \frac{Y(v)}{\mathcal{H}^*(v)}. \tag{3.119}$$

Deconvolution is often challenging since, at high frequencies, both $Y(v)$ and $\mathcal{H}^*(v)$ go to zero leaving the quotient highly uncertain. As a practical matter, the expression in (3.119) is often multiplied by an artificial transfer function to suppress high-frequency contributions to $X(v)$; thanks to the convolution theorem, this artificial transfer function, or "taper" as it is sometimes called, is equivalent to convolving the true result with a quantity that preserves the low- and medium-frequency components, but dramatically reduces the poorly resolved high frequencies in the problem.

One final related methodology deserves mention here. The *Wiener–Hopf method* is used in solving a class of partial differential equations as well as certain integral equations. It exploits the analytic properties of the various quantities employed in the complex plane. This methodology is presented in Mathews and Walker (1970). Chapter 8 of Morse and Feshbach (1999) provides a more detailed derivation of the method, while sections 9.6 and 18.4 of Stone and Goldbart (2009) provide derivations in different kinds of applications.

Having introduced Fourier series, Fourier integrals and transforms, some common examples, the autocorrelation function, Parseval's theorem, and convolutions, we turn now to issues emerging from band-limited processes, that is, those where no information is present outside a certain range of frequencies, time sampling of data, and efforts to approximate the evaluation of power spectra.

3.3.4   Sampling Theorem, Aliasing, and Approximation Methods

We wish to expand upon our treatment of deterministic problems to also include stochastic ones. In particular, most of the focus in the latter is upon the power spectrum as well as the autocorrelation function. As before, the pair

$$\rho(t) = \int_{-\infty}^{\infty} S(\nu) \exp(-2\pi i \nu t) \, d\nu$$

$$S(\nu) = \int_{-\infty}^{\infty} \rho(t) \exp(+2\pi i \nu t) \, dt \qquad (3.109)$$

remain valid. What is different now is the meaning of the autocorrelation function. In the deterministic case, we observed that

$$\rho(t) = \int_{-\infty}^{\infty} x(t + t') x^*(t') \, dt'. \qquad (3.111)$$

In the stochastic case, we define the autocorrelation function according to

$$\rho(t) = \langle x(t + t') x^*(t') \rangle. \qquad (3.120)$$

Here, the $\langle \cdots \rangle$ operator signifies that we are employing the expectation value, effectively the average, of many realizations of this random process. In chapter 5, we will address the issue of expectation values for random processes. Bracewell (2000) also addresses briefly the issue of estimating power spectra from random-process-associated data. We observe that this expression also depends upon what has become our benchmark time t'. We will assume that our process is *stationary*, that is, the result is independent of t'. In practice, practitioners sometimes assume that this result can be approximated by taking the mean value of such terms in a given realization of the process, that is, via averaging the products obtained from sampled values with a time lag between samples of t. We now return to (3.109), which remains valid in both cases.

Let us assume that $S(v)$, for either a deterministic or a stochastic signal, is band limited to the frequency interval $[-v_N, v_N]$, where v_N is called the *Nyquist frequency*, that is,

$$S(v) = 0 \quad \text{if } |v| > v_N. \tag{3.121}$$

The power spectrum, provided that it is well behaved, can be expressed as a linear combination with undetermined coefficients ρ_n of an infinite basis set, that is,

$$S(v) = \begin{cases} \Delta t \sum_{-\infty}^{\infty} \rho_n \exp(2\pi i n \Delta t v) & \text{if } |v| \leqslant v_N, \\ 0 & \text{if } |v| > v_N, \end{cases} \tag{3.122}$$

where

$$\Delta t = \frac{1}{B}, \tag{3.123}$$

where B is the bandwidth $2v_N$. We introduce this expression for the power spectrum into (3.109) and obtain

$$\rho(t) = \sum_{n=-\infty}^{\infty} \rho_n \frac{\sin[2\pi v_N (n\Delta t - t)]}{[2\pi v_N (n\Delta t - t)]}$$

$$= \sum_{n=-\infty}^{\infty} \rho_n \, \text{sinc} \, [2\pi v_N (n\Delta t - t)]. \tag{3.124}$$

When we seek to evaluate this expression at times t that are integer multiples of Δt, we observe that

$$\rho_n = \rho(\Delta t n), \tag{3.125}$$

thereby identifying that the unknown coefficients ρ_n are simply the sampled values of the autocorrelation function. The expression (3.124) is referred to as the *Whittaker interpolation formula*. Importantly, this equation demonstrates that $\rho(t)$ may be *exactly* determined at any time from a complete set of sampled autocorrelation function values. Moreover, the power spectrum is also exactly determined. This result, which is known as the *sampling theorem*, shows that no information about the power spectrum or the autocorrelation function is lost by sampling at intervals less than or equal to $(2\nu_N)^{-1}$.

Having proven the sampling theorem, it is instructive to see how sampling insufficiently frequently, that is, with $\Delta t > (2\nu_N)^{-1}$, results in an incorrect answer. Figure 3.4 demonstrates the phenomenon of *aliasing* (Blackman and Tukey, 1958).

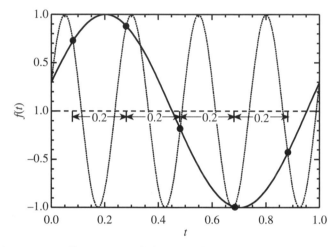

Figure 3.4. Illustration of aliasing: the inability to distinguish between two sinusoids fitting the observations.

Here we are looking at a one-second interval of time, and we took five samples, separated by $\Delta t = 0.2$ (four of which are identified using filled circles), in this diagram. Superposed on this diagram are two sinusoids, one at a frequency of 1 Hz and the other at 4 Hz. Since our sampling rate is 5 Hz, that is, the reciprocal of a 0.2-second sampling interval, what aliasing does is shift by ±5 Hz the true frequency. Thus, as our 4-Hz sinusoid has ±4-Hz components, a shift by ±5 Hz makes the 4-Hz sinusoid appear as though it were a $5 - 4 = 1$-Hz sinusoid. The plot makes the potential for this misidentification clear. Another familiar example of aliasing arises in television images of spoked wagon wheels turning on a Conestoga wagon in the seemingly wrong direction. The sampling rate associated with the number of images captured per second coupled with the rotation rate of the spokes combine to present this illusion: The spoke that appears to be moving in the wrong direction is in fact a different spoke in each of the successive images. The critical point to remember here is that information lost due to aliasing can never be recovered. Oversampling of data or the use of filters to further reduce the bandwidth are the only possible answers.

While our discussion and derivation has focused on continuously evolving variables, we wish to explore how we can effectively discretize our observations for band-limited signals. In order to do this, we utilize the sampling theorem so that we can exactly determine the Fourier transform and the power spectrum using appropriately sampled data. However, we also need to establish how we can make this procedure conform with a process where the band-limited signal becomes negligibly small after some time T or where the signal is periodic with period T. In principle, we could employ the Whittaker interpolation function to evaluate $\rho(t)$ or, for deterministic processes, $x(t)$. Our concern here is to calculate the relevant Fourier transform. Suppose, for the deterministic case, that $x(t)$ vanishes for $t < 0$ and essentially vanishes for $t > T$. In essence, we are concerned with the "energy" present in the signal; we select T so that almost all of the energy is in that time interval, say $\int_0^T |x(t)|^2 \, dt < \infty$, with essentially none present for $t < 0$ and $t > T$. In practical terms, this should be quantified. Depending upon the problem at hand, if 99% of the energy resides in $[0, T]$, then that criterion might be sufficient, or another criterion, such as 99.9%.

Since the signal is assumed to be band limited with the sampling interval $\Delta t \leqslant 1/B = 1/2\nu_N$, then we can write as before

$$f(\nu) = \begin{cases} \Delta t \displaystyle\sum_0^{N-1} x_n \exp(2\pi i n \Delta t \nu) & \text{if } |\nu| \leqslant \nu_N, \\ 0 & \text{if } |\nu| > \nu_N, \end{cases} \tag{3.126}$$

where $x_n = x(n\Delta t)$ and where $N \geqslant T/\Delta t$. (Note that a signal cannot in general be simultaneously band limited and time limited. However, if the energy resident outside the specified time interval is sufficiently small as discussed above, then this is a good approximation.) Suppose, now, that we seek the Fourier transform at discrete frequencies. The lowest frequency contribution that we can expect to find is $1/T$, corresponding to one event over the length of the observations. Hence, a reasonable choice of discrete frequencies would be multiples of that, that is, $n/T \approx n/N\Delta t$, so that

$$-\nu_N \leqslant \frac{n}{N\Delta t} \leqslant \nu_N$$

$$-\frac{1}{2\Delta t} \leqslant \frac{n}{N\Delta t} \leqslant \frac{1}{2\Delta t}$$

$$-\frac{N}{2} \leqslant n \leqslant \frac{N}{2}. \tag{3.127}$$

Accordingly, we will now define

$$f_m = \Delta t \sum_{n=0}^{N-1} x_n \exp\left(2\pi\mathrm{i}\frac{mn}{N}\right). \tag{3.128}$$

We observe that this formula is periodic over m with a period of N. While the meaning of this formula is clear given the bounds (3.3.4), it is convenient computationally to employ m in the same range as n, that is, $0 \leqslant m \leqslant N - 1$, and then shift the results for $m > N/2$ by N units so that the transform obtained is symmetric about the frequency origin. We now observe that

$$\sum_{m=0}^{N-1} f_m \exp\left(-2\pi\mathrm{i}\frac{m\ell}{N}\right) = \Delta t \sum_{m=0}^{N-1}\sum_{n=0}^{N-1} x_n \exp\left[2\pi\mathrm{i}\frac{m(n-\ell)}{N}\right]$$

$$= \Delta t \sum_{n=0}^{N-1} x_n N\delta_{n\ell}$$

$$= N\Delta t x_\ell. \tag{3.129}$$

Accordingly, it is conventional to write

$$x_\ell = \frac{1}{N\Delta t} \sum_{m=0}^{N-1} f_m \exp\left(-2\pi\mathrm{i}\frac{m\ell}{N}\right). \tag{3.130}$$

Taken together, Eqs. (3.128) and (3.130) are regarded as a *discrete Fourier transform (DFT)* pair.

From another perspective, we can visualize these two expressions presenting from an operational standpoint as an N-dimensional complex vector of f-components or of x-components being formed from a matrix product with the complementary vector. It is particularly instructive to identify the structure of that matrix. For example, consider $N = 8$ and employ (3.128) to compute the f_m transform terms. The matrix includes complex elements that are powers of

$$W \equiv \exp\left(\frac{2\pi i}{8}\right), \tag{3.131}$$

since $N = 8$. The matrix has the form

$$\begin{bmatrix} W^0 & W^0 & W^0 & W^0 & W^0 & W^0 & W^0 & W^0 \\ W^0 & W^1 & W^2 & W^3 & W^4 & W^5 & W^6 & W^7 \\ W^0 & W^2 & W^4 & W^6 & W^8 & W^{10} & W^{12} & W^{14} \\ W^0 & W^3 & W^6 & W^9 & W^{12} & W^{15} & W^{18} & W^{21} \\ W^0 & W^4 & W^8 & W^{12} & W^{16} & W^{20} & W^{24} & W^{28} \\ W^0 & W^5 & W^{10} & W^{15} & W^{20} & W^{25} & W^{30} & W^{35} \\ W^0 & W^6 & W^{12} & W^{18} & W^{24} & W^{30} & W^{36} & W^{42} \\ W^0 & W^7 & W^{14} & W^{21} & W^{28} & W^{35} & W^{42} & W^{49} \end{bmatrix}. \tag{3.132}$$

How can this be simplified?

3.3.5 Fast Fourier Transform

Cooley and Tukey (1965) discovered a remarkable scheme for simplifying a problem that fundamentally involved N^2 complex multiplications. Cooley et al. (1969) discussed further this discovery and how it could be simplified. Bracewell (2000), Brigham (1988), Nussbaumer (1982), and Walker (1988) provide very useful treatments of the *fast Fourier transform (FFT)*, and how it can be adapted to solving many problems. Importantly, as we shall see shortly, it is basically *identical* to the DFT: It simply provides a mechanism for performing the matrix multiply that we described in an extraordinarily simplified fashion.

Cooley and Tukey (1965) observed two fundamental features ensconced in the algebra. First, in computing mn/N, only the fractional part of the product of m with n contributed to the outcome since the integer portion simply resulted in rotations by some multiple of 2π in the phase of the complex arithmetic. Second, they identified a remarkable feature of binary arithmetic that would emerge if N were an integer power, say M, of 2, that is,

$$N = 2^M. \tag{3.133}$$

In practical terms, if N is not already a power of 2, then one can simply add additional components to the x-vector that are set to zero: They will not contribute to the computed f-vector and not otherwise cause difficulties. Improper use of FFT methods can create aliasing, for example, if sufficient care is not taken. The references cited above can help users of the FFT avoid such pitfalls. In particular, they observed that representing both m and n in binary arithmetic resulted in their product having a very special algebraic form involving M terms, and that performing the summations involved using individual binary or bit positions in the address of the vector components allowed for a dramatic simplification of the matrices involved. This can be seen by reviewing carefully Cooley and Tukey (1965) and reconstructing the algorithm that they developed. Indeed, Cooley and Tukey developed a scheme for eliminating all unnecessary or duplicated computations and produced an algorithm that is *irreducible*. Figure 3.5 highlights the intricacies that emerge in addressing and combining terms when performing a $32 = 2^5$ element FFT. We observe that the algebra couples elements that are separated by $1 = 2^0$ units, $2 = 2^1$ units, and so on, through $16 = 2^4$ units. This so-called "butterfly diagram" highlights the configurational complexity of the problem, which can be programmed nevertheless very efficiently, but does not clearly show *why* the FFT dramatically reduces the computational labor involved.

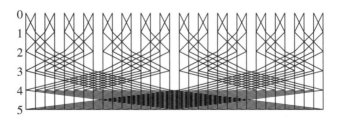

Figure 3.5. Data-flow butterfly diagram for a classical FFT where the vertical lines correspond to the number of elements, $N = 32$ in this case, in the time series. Each step proceeds to downward, ending in a node where two lines have been combined into one, via an addition and multiplication. The output emergent from the right-hand side corresponds with the 32 associated frequencies in a "bit-reversed array" in terms of the locations in memory of the results.

Gentleman (1968) provided a beautiful illustration of where this efficiency derives by using an $N = 8 = 2^3$ FFT working with the matrix (3.132), which we now present in simplified form and continue from there:

$$
\begin{bmatrix}
1 & 1 & 1 & 1 & 1 & 1 & 1 & 1 \\
1 & W & W^2 & W^3 & -1 & -W & -W^2 & -W^3 \\
1 & W^2 & -1 & -W^2 & 1 & W^2 & -1 & -W^2 \\
1 & W^3 & -W^2 & W & -1 & -W^3 & W^2 & -W \\
1 & -1 & 1 & -1 & 1 & -1 & 1 & -1 \\
1 & -W & W^2 & -W^3 & -1 & W & -W^2 & W^3 \\
1 & -W^2 & -1 & W^2 & 1 & -W^2 & -1 & W^2 \\
1 & -W^3 & -W^2 & -W & -1 & W^3 & W^2 & W
\end{bmatrix}
$$

$$
=
\begin{bmatrix}
1 & 0 & 0 & 0 & 1 & 0 & 0 & 0 \\
0 & 1 & 0 & 0 & 0 & W & 0 & 0 \\
0 & 0 & 1 & 0 & 0 & 0 & W^2 & 0 \\
0 & 0 & 0 & 1 & 0 & 0 & 0 & W^3 \\
1 & 0 & 0 & 0 & -1 & 0 & 0 & 0 \\
0 & 1 & 0 & 0 & 0 & -W & 0 & 0 \\
0 & 0 & 1 & 0 & 0 & 0 & -W^2 & 0 \\
0 & 0 & 0 & 1 & 0 & 0 & 0 & -W^3
\end{bmatrix}
$$

$$
\times
\begin{bmatrix}
1 & 0 & 1 & 0 & 0 & 0 & 0 & 0 \\
0 & 1 & 0 & 0 & 0 & W^2 & 0 & 0 \\
1 & 0 & -1 & 0 & 0 & 0 & 0 & 0 \\
0 & 1 & 0 & -W^2 & 0 & 0 & 0 & 0 \\
0 & 0 & 0 & 0 & 1 & 0 & 1 & 0 \\
0 & 0 & 0 & 0 & 0 & 1 & 0 & W^2 \\
0 & 0 & 0 & 0 & 1 & 0 & -1 & 0 \\
0 & 0 & 0 & 0 & 0 & 1 & 0 & -W^2
\end{bmatrix}
$$

$$
\times
\begin{bmatrix}
1 & 1 & 0 & 0 & 0 & 0 & 0 & 0 \\
1 & -1 & 0 & 0 & 0 & 0 & 0 & 0 \\
0 & 0 & 1 & 1 & 0 & 0 & 0 & 0 \\
0 & 0 & 1 & -1 & 0 & 0 & 0 & 0 \\
0 & 0 & 0 & 0 & 1 & 1 & 0 & 0 \\
0 & 0 & 0 & 0 & 1 & -1 & 0 & 0 \\
0 & 0 & 0 & 0 & 0 & 0 & 1 & 1 \\
0 & 0 & 0 & 0 & 0 & 0 & 1 & -1
\end{bmatrix}
$$

$$
\times
\begin{bmatrix}
1 & 0 & 0 & 0 & 0 & 0 & 0 & 0 \\
0 & 0 & 0 & 0 & 1 & 0 & 0 & 0 \\
0 & 0 & 1 & 0 & 0 & 0 & 0 & 0 \\
0 & 0 & 0 & 0 & 0 & 0 & 1 & 0 \\
0 & 1 & 0 & 0 & 0 & 0 & 0 & 0 \\
0 & 0 & 0 & 0 & 0 & 1 & 0 & 0 \\
0 & 0 & 0 & 1 & 0 & 0 & 0 & 0 \\
0 & 0 & 0 & 0 & 0 & 0 & 0 & 1
\end{bmatrix} . \tag{3.134}
$$

In particular, he showed that this 8×8 complex matrix could be reduced to a product of three matrices, since $8 = 2^3$, plus a fourth to accommodate the assignment of proper memory locations to the results, matrices that had, at most, one or two nontrivial components per row. We now show this matrix, simplified slightly in form, and then its deconstruction into a product of three sparse matrices, each with only $2N$ nonvanishing elements, and the rearrangement of terms.

Consider, for example, a situation where $M = 20$ so that $N = 1,048,576$ terms, essentially a million-point Fourier transform. The formal matrix multiplication involves N^2 terms, which is $\approx 10^{12}$, a prodigious amount of computation. However, the binary deconstruction of the matrix will result in M or 20 extremely sparse matrices plus a rearrangement matrix. The computational effort, then, involves $M = \log_2 N$ matrix-multiply operations with $2N$ products per multiply, only N of which require a complex multiplication. Hence, the FFT involves $N \log_2 N$ or MN complex multiplies, which, for $N = 2^{20}$, reduces to $\approx 2 \times 10^7$ operations, a reduction in effort of five orders of magnitude.

The FFT has become one of the most commonly employed algorithms in geophysics and related disciplines. The key to success in applying the FFT is to remember that it is identical to the DFT when $N = 2^M$ and dramatically more efficient. At the same, we must remember that it is only as accurate as the DFT, so that substantial care must be taken in its construction and application. It is useful to note that the FFT can also be employed in applications to the Chebyshev polynomials and that there also exists a "fast" Legendre transform; Glatzmaier (2013) provides some insight into their implementation. We now turn our attention to inverse theory and to other integral transform problems that emerge, especially in geophysical and other tomographic applications, namely, the Abel, Radon, and Herglotz–Wiechert transforms.

3.4 Inverse Theory, Calculus of Variations, and Integral Equations

Inverse theory (Parker, 1977, 1994) is an important arena in geophysics and elsewhere. We will begin by providing a brief introduction to the subject. We will employ the notation of Parker

(1977), who introduced the problem via the linear integral equation or integral transform

$$e(x) = \int_I G(x, y) m(y) \, dy, \qquad (3.135)$$

where m is the unknown function we are seeking, e a function representing "observations," G a kernel derived from theory, and I a real interval. We have already encountered several such examples, devoting substantial attention to the Fourier transform wherein

$$G(x, y) = \exp(2\pi i x y). \qquad (3.136)$$

The Laplace and Hankel transforms also have this flavor. We will focus here on three related problems of this type that have important applications to astronomy and medicine, as well as geophysics, namely, the Abel, Radon, and Herglotz–Wiechert transforms. However, we will begin by exploring an important subset of such problems that often emerge in experiments.

3.4.1 Linear Inverse Theory

Inverse theory is often employed in the context of experimental data where observations are only available for sampled values of x, say, x_i. For simplicity, we will express $G(x_i, y)$ as $G_i(y)$ and $e(x_i)$ as e_i, whereupon Eq. (3.135) becomes

$$e_i = \int_I G_i(y) m(y) \, dy. \qquad (3.137)$$

A familiar example of this kind of problem emerges when the G_i are (normalized) spherical harmonics, which we shall describe in greater depth in the following chapter. In that instance, the e_i are the spherical harmonic expansion coefficients, and knowledge of these immediately allows us to present an approximate solution for $m(y)$. We have already explored a special case in the context of the DFT and the FFT and their limitations via the sampling theorem earlier in this chapter. However, suppose we are presented with the more general circumstance where the kernel functions $G_i(y)$ are in some sense arbitrary, the outcome of a complicated observing process. To what extent can we employ the e_i to estimate $m(y)$, and can we identify the uncertainty in our estimate?

Backus and Gilbert (1968, 1970) were the first to systematize this process, and Parker (1994), especially, explored the process in a broader context. The primary focus of Backus and Gilbert was to identify and in some sense minimize the uncertainty in the estimate available for $m(y)$; although many researchers have employed their methodology to estimate the solution $m(y)$, this was never their intent. There is a very rich literature on this topic in applications to geophysics as well as to other disciplines, for example, Newman (1979b). We now turn our attention to some real-world problems that emerge when the unknown function is being integrated along some line of sight or, of special interest to seismology, along the path taken by a seismic disturbance.

3.4.2   Abel Transform

Many problems in geophysics and allied disciplines relate to the inversion of line-of-sight observations of environments that display circular symmetry, the so-called *Abel integral transform* problem (Bracewell, 2000; Sneddon, 1972). In Figure 3.6, we give a typical example where observations correspond to an integration along a line of sight, the y-axis for $-\infty < y < \infty$, of a distribution $f(x, y)$ as a function of x, which we will call $F(x)$.

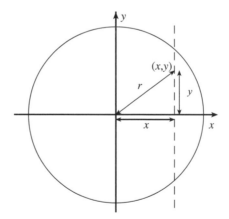

Figure 3.6.   Illustration of Abel transformation geometry.

Further, we assume for the moment that the source is circularly symmetric, that is, $f(x, y) \to f(r)$, thereby giving

$$F(x) = \int_{-\infty}^{\infty} f(x, y)\, dy$$

$$= 2 \int_{r=x}^{\infty} f(r) \frac{r\, dr}{\sqrt{r^2 - x^2}}, \tag{3.138}$$

where we have employed the usual Pythagorean identity $r^2 = x^2 + y^2$. The objective here is to find $f(r)$, given that $F(x)$ has been measured. Bracewell provides a comprehensive treatment and employs this notation, although many other notations appear in the literature (Sneddon, 1972). This integral equation appears in some surprising contexts (Newman, 1979a).

We multiply the latter quantity by a structurally similar quantity to define

$$
\begin{aligned}
\mathcal{F}(z) &= \int_{x=z}^{\infty} \frac{F(x)x\,dx}{\sqrt{x^2 - z^2}} \\
&= \int_{x=z}^{\infty} \frac{2x\,dx}{\sqrt{x^2 - z^2}} \left\{ \int_{r=x}^{\infty} \frac{f(r)r\,dr}{\sqrt{r^2 - x^2}} \right\} \\
&= \int_{r=z}^{\infty} f(r)r\,dr \left\{ \int_{x=z}^{r} \frac{2x\,dx}{\sqrt{x^2 - z^2}\sqrt{r^2 - x^2}} \right\}. \quad (3.139)
\end{aligned}
$$

We make the substitution $u = (x^2 - z^2)/(r^2 - z^2)$ and then employ $u = \sin^2 \theta$ to obtain

$$
\begin{aligned}
\mathcal{F}(z) &= \int_{r=z}^{\infty} f(r)r\,dr \int_0^1 \frac{du}{\sqrt{u}\sqrt{1-u}}, \\
&= \pi \int_{r=z}^{\infty} f(r)r\,dr. \quad (3.140)
\end{aligned}
$$

Finally, we observe that

$$
\frac{d\mathcal{F}(z)}{dz} = -\pi f(z)z. \quad (3.141)
$$

While formally correct, we must remember that $F(x)$ is generally obtained from observations and the calculation of $\mathcal{F}(z)$ involves an integral containing a seemingly singular integrand. Since the end result requires the calculation of a derivative, the outcome can be rather noisy.

Another approach to solving this problem emerges from taking the Fourier transform of $F(x)$, namely,

$$
\begin{aligned}
\int_{-\infty}^{\infty} & F(x)\exp(ikx)\,dx \\
&= \int_{-\infty}^{\infty} dx \exp(ikx) \left\{ 2 \int_{r=|x|}^{\infty} f(r) \frac{r\,dr}{\sqrt{r^2 - x^2}} \right\} \\
&= 2 \int_{r=0}^{\infty} f(r)r\,dr \int_{x=-r}^{r} \frac{\exp(ikx)\,dx}{\sqrt{r^2 - x^2}}. \quad (3.142)
\end{aligned}
$$

We make the substitution $x = r \sin \theta$, whereupon we obtain

$$\int_{-\infty}^{\infty} F(x) \exp(ikx) \, dx$$

$$= 2 \int_{r=0}^{\infty} f(r) r \, dr \left\{ \int_{\theta=-\pi/2}^{\pi/2} \exp(ikr \sin \theta) \, d\theta \right\}$$

$$= \int_{r=0}^{\infty} f(r) r \, dr J_0(kr), \qquad (3.143)$$

which we observe corresponds to the Fourier–Bessel transform (3.105) and (3.106).

3.4.3   Radon Transform

The general case of integrated line-of-sight observations forms the basis of computed tomography and is known as the *Radon transform*. Tomography exploits any kind of wave, such as light waves, that travels through a two-dimensional cross section of a material, where we are able to measure its amplitude upon exiting the material as a function of angle and the offset of that "ray" from the origin. The solution is detailed in Deans (2007) and Walker (1988), and is generally presented in the context of Hilbert transforms. In essence, this methodology relates effectively to rotating the perpendicular line or tangent line that connects the origin to the line of sight. However, there is a simpler way to perceive this problem if we express our unknown source $f(x, y)$ in polar coordinates, namely, $f(r, \theta)$, and employ Fourier series (3.100), namely,

$$f(r, \theta) = \sum_{-\infty}^{\infty} f_n(r) \exp(in\theta). \qquad (3.100)$$

Owing to the inherent linearity in the problem, the integrated line-of-sight quantity $F(x)$ can now be expressed as $F(x, \phi)$, where ϕ now relates to the angle between that tangent line and the x-axis. This function, too, can be expressed as a Fourier series, and the Abel transform that we have derived above can now be applied to each of the azimuthal terms that emerge in this expansion. Finally, three-dimensional tomographic images can be constructed by creating a stack of "slices" of two-dimensional profiles.

We now briefly review the *calculus of variations* inasmuch as it has an important role in the evaluation of integral quantities. We will then turn to a specialized but extremely important class of transforms that appears in seismology and emerges from minimizing the time taken for a seismic event to propagate from its source to its observer.

3.4.4 Calculus of Variations

In many circumstances, we calculate an integrated quantity I that depends upon the path taken—which may not be a straight line. Here, unlike the issues addressed earlier relating to contour integration, the path taken does matter. For example, we think of the shortest path between two points as a straight line. More detailed treatments of this problem can be found in Goldstein et al. (2002), Boas (2006), and Butkov (1968). Let us consider a path going from coordinate (x_1, y_1) to (x_2, y_2). We introduce a differential line element ds that describes, given dx and dy, the infinitesimal distance travelled, that is,

$$ds^2 = dx^2 + dy^2. \tag{3.144}$$

It follows that the distance travelled in going from one point to the other, if the path taken is specified by $y(x)$, is given by

$$I = \int_{(x_1, y_1)}^{(x_2, y_2)} ds = \int_{(x_1, y_1)}^{(x_2, y_2)} \sqrt{1 + \left[\frac{dy(x)}{dx}\right]^2} \, dx. \tag{3.145}$$

We now want to identify the conditions that render the path optimal by minimizing I. Suppose, now, that the integrand could be expressed as a function of $y(x)$ and its derivative $dy(x)/dx$; call it $L[y(x), dy(x)/dx]$. We have used the symbol L to denote the integrand, recalling the contributions made to this problem area by Lagrange. For the case of minimizing the path length in going from point 1 to point 2, we observe that

$$L\left[y(x), \frac{dy(x)}{dx}\right] = \sqrt{1 + \left[\frac{dy(x)}{dx}\right]^2}. \tag{3.146}$$

Let us identify the change in the integral I that emerges if the path now becomes $y(x) + \delta y(x)$ but begins and ends at the original specified points. (Here, we are employing δ to refer to an infinitesimal displacement, and it is not related to the Dirac δ function.) Thus, we obtain

$$\delta I = \int_{(x_1, y_1)}^{(x_2, y_2)} \left[\frac{\partial L}{\partial y} \delta y + \frac{\partial L}{\partial y'} \delta y'\right] dx, \tag{3.147}$$

where $y'(x) = dy(x)/dx$. We wish to employ integration by parts to render the second term as being expressed using δy; we note that

$$\delta y'(x)\, dx = d[\delta y(x)], \tag{3.148}$$

thereby allowing us to write

$$\delta I = \int_{(x_1, y_1)}^{(x_2, y_2)} \left[\frac{\partial L}{\partial y} - \frac{d}{dx}\left(\frac{\partial L}{\partial y'} \right) \right] \delta y(x)\, dx. \tag{3.149}$$

Since we want our path to be optimal, we require that δI vanish for any variation in the path $\delta y(x)$, whereupon we obtain the *Euler–Lagrange equation*

$$\frac{\partial L}{\partial y} = \frac{d}{dx}\left(\frac{\partial L}{\partial y'} \right). \tag{3.150}$$

In the minimum path-length problem (3.146), we observe that L has no y-dependence, thereby implying that $\partial L/\partial y'$, which is a function of y' alone, is a constant. This yields the result that y' is the constant (slope) that carries the solution from point 1 to point 2, that is,

$$y'(x) = \frac{y_2 - y_1}{x_2 - x_1}. \tag{3.151}$$

A very simple variant of this problem is to identify the path with the shortest travel time where $v(x)$ is the propagation speed. In that instance, we observe that

$$L\left[y(x), \frac{dy(x)}{dx} \right] = \frac{\sqrt{1 + (dy(x)/dx)^2}}{v(x)}. \tag{3.152}$$

We leave as an exercise the derivation of Snell's law for a medium that has the wave velocity change from v_1 to v_2 when the wave passes through a material interface x_m with $x_1 < x_m < x_2$. Snell's law resides at the heart of seismic travel-time analysis, which we now present.

3.4.5 Herglotz–Wiechert Travel-Time Transform

The determination of seismic velocity as a function of depth using first-arrivals travel time information is a mainstay of solid earth geophysics. Bullen (1963), Aki and Richards (2002), and Shearer (2009) provide incisive discussions of the topic and its limitations. Fundamentally, the methodology developed by

Herglotz and Wiechert based upon Fermat's principle wherein the path taken by a propagating wave, often called a "teleseismic ray" in seismology, is that which takes the least amount of time. The key ingredient here is Snell's law, which relates the angle of incidence to that of transmission in geometrical optics across an abrupt change in material properties.

We begin with the simplest situation of a spherically symmetric and homogeneous Earth with radius R where the seismic velocity is constant, say v, and traverses a distance D along a chord subtending an angle of Δ as in Figure 3.7.

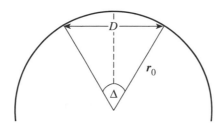

Figure 3.7. Illustration of teleseismic ray geometry.

It follows immediately that the travel time T is simply

$$T = \frac{D}{v} = \frac{2r_0}{v} \sin\left(\frac{\Delta}{2}\right), \qquad (3.153)$$

where r_0 is the Earth's radius. Since the seismic velocity of the Earth varies with radius, generally increasing with depth, we need to introduce a methodology to perform the inversion. Unlike the straight-line character depicted in Figure 3.7, the path taken by seismic waves will drop below to take advantage of higher seismic velocities, thereby shortening the first-arrival time. Complications emerge in the analysis of observational data owing to limited numbers of stations and the accuracy of first-arrival estimates, but also due to behaviors such as "shadowing" and "triplication" that emerge due to departures from monotonic increase in seismic speed with respect to depth.

Consider now Figure 3.8 (Bullen, 1963; Aki and Richards, 2002; Shearer, 2009) illustrating the role of Snell's law in a spherically stratified Earth. (In the mathematical limit, the "discontinuities" can disappear and the model describes a seismic velocity that varies smoothly with depth.)

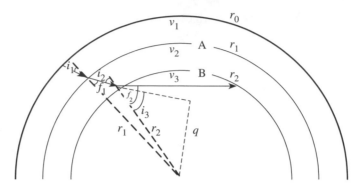

Figure 3.8. Illustration of layered geometry for Snell's law.

In particular, we apply Snell's law to the two discontinuities at A and B as shown in Figure 3.8 and observe

$$\frac{\sin i_1}{v_1} = \frac{\sin f_1}{v_2}$$

$$\frac{\sin i_2}{v_2} = \frac{\sin f_2}{v_3}, \tag{3.154}$$

and so on. However, we also note from the disposition of the two right-angled triangles that

$$q = r_1 \sin f_1 = r_2 \sin f_2, \tag{3.155}$$

and so on. Combining these expressions, we then observe that

$$\frac{r_1 \sin i_1}{v_1} = \frac{r_1 \sin f_1}{v_2} = \frac{r_2 \sin i_2}{v_2} = \frac{r_2 \sin f_2}{v_3}, \tag{3.156}$$

and so on. The outcome of this process in looking at any number of layers with increasing seismic velocity is that we observe a constant quantity p, called the *ray parameter*, defined by

$$p = \frac{r \sin i}{v}, \tag{3.157}$$

where i is the angle between the ray and the radial vector at any point, and we will presume that

$$\frac{dv}{dr} < \frac{v}{r}. \tag{3.158}$$

We must now establish a relationship between p and the travel-time period by considering two infinitesimally displaced rays as shown in Figure 3.9.

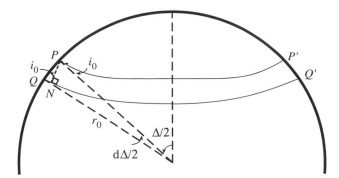

Figure 3.9. Illustration of adjacent teleseismic geometry.

Our purpose here is to derive a differential equation governing the variation of the ray parameter with respect to the corresponding depth of greatest penetration. From Figure 3.9 it follows that the line segments QN and PQ are related by

$$QN = PQ \sin i_0 = \tfrac{1}{2} r_0 \, d\Delta \sin i_0. \qquad (3.159)$$

Since the two teleseisms experience a difference in arrival time dT of $2 \times QN/v$, we have

$$\frac{dT}{d\Delta} = \frac{r_0 \sin i_0}{v_1} = p. \qquad (3.160)$$

Given this identity, it is convenient to define the ray parameter as a function of the angle Δ, namely,

$$p(\Delta) = \frac{dT(\Delta)}{d\Delta}, \qquad (3.161)$$

and take note that this quantity will not be accurately known owing to noise in the measurements. Later in this derivation, we will regard Δ as a function of p. In operational terms, this can be accomplished computationally using a variety of curve-fitting techniques, such as cubic splines. Using polar geometry, we can consider a line element ds along a teleseism that satisfies

$$(ds)^2 = (dr)^2 + (r \, d\theta)^2, \qquad (3.162)$$

whereupon we observe that

$$p = \frac{r}{v} \cdot \frac{r \, d\theta}{ds} \qquad (3.163)$$

along the ray path. It is important to note that p is now a function of r, that is, $r/v(r)$, and that it increases as r increases. We eliminate ds between the latter two equations and observe that

$$\left(\frac{r^2 \, d\theta}{vp}\right)^2 = (dr)^2 + (r \, d\theta)^2. \tag{3.164}$$

It is convenient to introduce a factor $\eta = r/v$, whereupon we obtain

$$\frac{d\theta}{dr} = \frac{p}{r(\eta^2 - p^2)^{1/2}}, \tag{3.165}$$

and integrate from the deepest point of penetration of the ray, call it r', to the surface r_0 to obtain

$$\tfrac{1}{2}\Delta = \int_{r'}^{r_0} \frac{p \, dr''}{r''[\eta^2 - p^2]^{1/2}}. \tag{3.166}$$

Here, $\eta(r'')$ varies while p is held fixed at $p(r')$; for convenience, we will replace $\eta(r'')$ by η_1 and $\eta(r_0)$ by η_0. This expression can now be converted into a form amenable to the Abel transform inversion (Aki and Richards, 2002; Shearer, 2009).

We take the limits of integration as being η_0, at the surface r, and $\eta = p$, corresponding to the deepest point of the ray path. We can write

$$\Delta(p) = \int_{p}^{\eta_0} \frac{2p}{r''[\eta^2 - p^2]^{1/2}} \frac{dr}{d\eta} \, d\eta, \tag{3.167}$$

where we regard η as being implicitly a function of r'' and, conversely, r'' as being implicitly a function of η. Following the same integration technique employed in the Abel transform, we evaluate

$$\int_{\eta_1}^{\eta_0} \frac{\Delta(p) \, dp}{(p^2 - \eta_1^2)^{1/2}}$$
$$= \int_{\eta_1}^{\eta_0} \left\{ \int_{p}^{\eta_0} \frac{2p}{r[(p^2 - \eta_1^2)(\eta_0^2 - p^2)]^{1/2}} \frac{dr}{d\eta} \, d\eta \right\} dp. \tag{3.168}$$

As in the Abel case, we employ the identity

$$\int_{\eta_1}^{\eta_0} \frac{p \, dp}{[(p^2 - \eta_1^2)(\eta_0^2 - p^2)]^{1/2}} = \frac{\pi}{2}. \tag{3.169}$$

Thus, the right-hand side of (3.168) becomes

$$\pi \int_{\eta_1}^{\eta_0} \frac{1}{r} \frac{dr}{d\eta} \, d\eta = \pi \ln\left(\frac{r_0}{r_1}\right). \tag{3.170}$$

To address the right-hand side, we employ the substitution

$$p = \eta_1 \cosh(u), \tag{3.171}$$

and thereby obtain for the left-hand side that

$$\int_{\eta_1}^{\eta_0} \frac{\Delta(p)\,\mathrm{d}p}{(p^2 - \eta_1^2)^{1/2}} = \int_0^{\cosh^{-1}(p/\eta_1)} \Delta(u)\,\mathrm{d}u, \qquad (3.172)$$

with $u = \cosh^{-1}(p/\eta_1)$. Then, we integrate by parts to obtain

$$\int_0^{\cosh^{-1}(p/\eta_1)} \Delta(u)\,\mathrm{d}u$$

$$= \left[\Delta(p) \cosh^{-1}\left(\frac{p}{\eta}\right) \right]_{p=\eta_1}^{\eta_0} - \int_{\eta_1}^{\eta_0} \frac{\mathrm{d}\Delta}{\mathrm{d}p} \cosh^{-1}\left(\frac{p}{\eta_1}\right) \mathrm{d}p. \quad (3.173)$$

The leading term on the left-hand side vanishes since $\Delta = 0$ when $p = \eta_0$ and $\cosh^{-1}(p/\eta_1) = 0$ when $p = \eta_1$. Therefore, we are left with

$$\int_0^{\Delta_1} \cosh^{-1}\left[\frac{p(\Delta)}{\eta_1} \right] \mathrm{d}\Delta = \pi \ln\left(\frac{r_0}{r_1}\right). \qquad (3.174)$$

Since η has the value of r/v at the point of deepest penetration while traveling a distance Δ, it is convenient to replace η_1 with p_1. Earlier, we tabulated $p(\Delta)$ from the observed travel times $T(\Delta)$. Accordingly, by using numerical interpolation of tabulated $\Delta - T$ values and their associated derivative, namely, $p(\Delta)$, we can numerically integrate using tabular interpolation the integral appearing on the left-hand side for each value of Δ_1 (corresponding to p_1). From these tabulated results, we can then establish the value of r_1 associated with Δ_1, the depth of deepest penetration. Moreover, since we already know $p(\Delta_1)$, we can then establish the seismic velocity $v(r_1)$ at that depth. Shearer (2009) provides a detailed description of how to implement this scheme.

While this process is more involved than the usual one associated with the Abel transform and similarly is prone to observational errors and the need to take numerical estimates of derivatives, as well as physical effects such as shadowing and triplication, we can combine the observations routinely made in assessing travel times and produce an estimate of seismic velocity as a function of depth. Seismic tomography has joined the ranks of computed tomography in providing important insights into the interior of the Earth.

This concludes our discussion of integral transforms and some of the many guises they assume in a broad array of applications. We return now to more classical themes in the application of mathematical methods to the fundamental types of equations encountered in geophysics and other disciplines.

3.5 Exercises

1. Evaluate the fourth moment of a Gaussian using the methods we have derived, that is, evaluate

$$\int_{-\infty}^{\infty} (x - \mu)^4 \mathcal{N}(x; \mu, \sigma) \, dx.$$

2. Verify the identity $\Gamma(z + 1) = z\Gamma(z)$ using integration by parts.

3. Verify that $\Gamma(1/2) = \sqrt{\pi}$.

4. By keeping linear and quadratic terms in the Taylor series expansion for $f(x)$ in the neighborhood of x_0, show that the error in this expansion is approximately

$$\frac{f''(x_0)}{2g''(x_0)} \exp[-g(x_0)] \sqrt{\frac{2\pi}{g''(x_0)}}.$$

5. Look up in a standard reference text the error function or $\mathrm{erf}(x)$ and its "complement" $\mathrm{erfc}(x)$ and provide their definitions and describe explicitly how each is related to $\mathcal{N}(x; \mu, \sigma)$.

6. As a practical test, use your choice of Matlab, Mathematica, or Maple to evaluate to six significant digits 50! using the exact expression, that is, $50 \cdot 49 \cdots 2 \cdot 1$ and the more accurate version of Stirling's approximation (3.30).

7. Using the techniques developed earlier, calculate $\Xi_3(1)$ using the power series method and $\Xi_3(10)$ using the asymptotic series method. In each instance, derive the series to incorporate five terms, and estimate the error in each.

8. Using the definition of elliptic integrals, evaluate

$$\int_0^{\pi/4} \frac{d\varphi}{\sqrt{12 + 8\cos^2 \varphi}}$$

in terms of the appropriate form for an incomplete elliptic integral. Finally, use Matlab, Mathematica, or Maple to evaluate this quantity numerically.

9. Consider the integral

$$\int_{-\infty}^{\infty} \frac{dx}{1 + x^2}.$$

(a) Evaluate this integral using elementary trigonometric identities.

(b) You will now employ contour integration techniques to evaluate this integral along the contour shown.

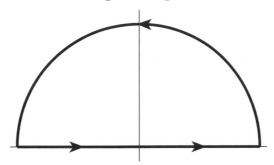

What this diagram illustrates is that the contour has two parts. The lower segment corresponds to the real axis while the upper one corresponds to a semicircle that connects with the line at arbitrarily large positive and negative values of x. First, you must show that the infinite arc does not contribute to the contour by employing a semicircle with radius $R \to \infty$. In so doing, you will have shown that the only contribution to the integral here is from the line segment along the real axis, and that the value of the integral can be obtained by calculating the residue, which you should now do.

10. Consider the following time periodic functions:

(a)

$$x(t) = \begin{cases} t & \text{if } 0 \leqslant t < \frac{1}{2}, \\ -t & \text{if } -\frac{1}{2} \leqslant t < 0. \end{cases}$$

Calculate its Fourier series representation.

(b)

$$x(t) = \begin{cases} 1 & \text{if } 0 \leqslant t < \frac{1}{2}, \\ -1 & \text{if } -\frac{1}{2} \leqslant t < 0. \end{cases}$$

Calculate its Fourier series representation.

(c) How are the series for the two problems related? Note that the derivative of the second case is equal to that of the first case.

(d) What is the derivative of $x(t)$ in the first part? How is it related to a δ function?

11. For the boxcar function (3.83), calculate the autocorrelation function corresponding to it. Plot the result.

12. Consider a normalized Gaussian distribution

$$x(t) = \frac{1}{\sqrt{2\pi\sigma^2}} \exp\left(-\frac{x^2}{2\sigma^2}\right).$$

(a) Calculate according to (3.111) its autocorrelation function.

(b) Now, calculate its Fourier transform $f(v)$.

(c) Use the convolution theorem to obtain the autocorrelation function from $f(v)$ by inverting the associated Fourier transform.

13. Consider the function, for $\tau > 0$ and all time t,

$$x(t) = \exp\left(-\frac{|t|}{\tau}\right).$$

(a) Calculate its Fourier transform $f(v)$, and identify the location and nature of its poles.

(b) Calculate the inverse transform for $t > 0$ using an appropriate contour. Explain

(c) Similarly, repeat the calculation for $t < 0$. How do the contours differ from each other?

14. Consider the Gaussian function in problem 3 with $\sigma = 1$ and its Fourier transform.

(a) From a practical standpoint, develop a criterion for the "effective bandwidth" and, hence, the sampling interval Δt.

(b) Similarly, over what range of time should you take samples? [Hint: Think in terms of an "energy criterion." How does Parseval's theorem have a role in this?]

(c) Consequently, what number N of samples should be taken?

15. Select the smallest integer M so that $2^M > N$. Compare the relative amount of effort, in terms of "multiplication counts," of performing a DFT and an FFT.

16. Suppose you have a circularly symmetric spatial distribution of mass for which integrated line-of-sight observations are available, as associated with the Abel transform. Let us assume that the radial distribution has the form

$$f(r) = \exp(-r^2).$$

It is noteworthy in many astronomical as well as exoplanetary observations that this kind of problem emerges.

(a) Calculate the line-of-site distribution $F(x)$.

(b) Then, calculate the quantity $\mathcal{F}(z)$ associated with its inversion. Using this result, take its derivative in order to obtain the original distribution $f(z)$.

17. Suppose, in the x-y plane you have one medium for $y > 0$ with propagation speed v_1 and another medium for $y < 0$ with propagation speed $v_2 > v_1$.

(a) By minimizing the travel time for a signal, show that

$$\frac{\sin \theta_1}{v_1} = \frac{\sin \theta_2}{v_2}.$$

(b) Suppose the angle of incidence θ_1 satisfies

$$\frac{v_2 \sin \theta_1}{v_1} > 1.$$

Physically, explain what is happening here.

18. An interesting atmospheric phenomenon illustrates a well-known seismological phenomenon, that of *shadowing*. This describes phenomena where the signal cannot be observed by an observer due to the curvature in the ray path. In an adiabatic atmosphere near the surface, the temperature gradient is negative and uniform (around $-6°$/km). It is sufficient for our purposes to assume that the Earth is (locally) flat. However, it should be noted that the shadowing phenomenon that occurs here is an outcome of the Earth's curvature. For convenience, we can assume that the sound speed v as a function of altitude satisfies

$$v(z) = \sqrt{a - bz},$$

where a and b are positive constants. Suppose you see a distant lightning flash relatively close to the ground—but you do not hear it. Further suppose that the source of the lightning is at $(0, H)$ in the x-z plane.

(a) Apply Fermat's principle of least time and assume that you, the observer, are at a position $(D, 0)$. Show that the integral that must be minimized has the form

$$I = \int \sqrt{\frac{1 + z'^2(x)}{a - bz(x)}} \, dx.$$

(b) Show that the Euler–Lagrange equation for this system is

$$\frac{d}{dx}\left(\frac{\partial \mathcal{L}}{\partial z'}\right) = \frac{\partial \mathcal{L}}{\partial z},$$

where $\mathcal{L}$ is the integrand in the previous equation. What is the second-order differential equation for $z(x)$ that must be satisfied?

19. Instead of using the standard derivation of the Herglotz–Wiechert travel time inversion, consider using polar coordinates and the calculus of variations where the seismic velocity $v(r)$ is to be determined. Formulate the variational problem that must be solved in minimizing the travel time taken by a seismic wave in traveling an angular distance Δ from source to observer. Once you have obtained the associated Lagrangian, solve the equation that results and show that it is equivalent to (3.157).

CHAPTER FOUR

Partial Differential Equations of Mathematical Geophysics

Partial differential equations have been the mainstream of theoretical geophysics for many decades. For example, traditional treatments of the subject (Stacey and Davis, 2008) effectively subdivide the field according to the nature of the underlying problem as well as the methodologies employed for solving them. Here, we will review the classification of partial differential equations and their boundary conditions, and then proceed to review the wave equation in one dimension, introduce some of the fundamental equations associated with geophysical fluids and the forces that drive them, explore the kinds of potential equations that can occur, and advance to explore thermal diffusion and the wave equation in three dimensions. We then return to the gravitational potential and Green's function methods of solution. After identifying the nature of diffusion and dispersion, we explore sound waves and elementary perturbation theory. We then advance to nonlinear partial differential equations in geophysics, including those due to Burgers in the context of solitary waves and Korteweg–de Vries in the context of solitons. We conclude the chapter with the nature of self-similarity, scaling phenomena, and Kolmogorov's statistical theory of homogeneous, isotropic turbulence and the role of complexity.

4.1 Introduction to Partial Differential Equations

We begin by providing the classification of partial differential types and boundary condition types, and then introduce the wave equation in one dimension, some elementary ideas in fluid flow, and the diffusion equation.

4.1.1 Classification of Partial Differential Equations and Boundary Condition Types

There are three major types of partial differential equations, as well as mixtures of these types, and appreciating the distinction among them is critical to understanding their behavior. The differences that manifest are intimately tied to geometry and the local behavior seen in these equation types. The text by Courant and Hilbert (1962) is an authoritative resource for this topic, but we present here an abbreviated treatment of the classification as well as the different types of boundary conditions.

The three types of partial differential equations are known as *hyperbolic*, *elliptic*, and *parabolic*. Consider a so-called "quasi-linear" first-order equation, with independent variables x and y, for the quantity $u(x, y)$, namely,

$$a\frac{\partial u}{\partial x} + b\frac{\partial u}{\partial y} = c, \tag{4.1}$$

which is sometimes written

$$ap + bq = c, \tag{4.2}$$

where a, b, and c are functions of x, y, and u, but not of its partial derivatives, sometimes designated by $p = \partial u/\partial x$ and $q = \partial u/\partial y$. Note that the variables need not designate spatial quantities; for example, y can correspond to time.

We need to explore how this equation affects the behavior of u on a line C in the x-y plane. In particular, we wish to understand how that information can be used to uniquely determine u elsewhere. Let us imagine, now, that $u(x, y)$ describes a surface and thereby constitutes a third dimension (analogous to z). Then, a normal to that surface at (x, y) has direction cosines p, q, and -1 since we can write

$$\frac{\partial u}{\partial x}\,\mathrm{d}x + \frac{\partial u}{\partial y}\,\mathrm{d}y - \mathrm{d}u = 0, \tag{4.3}$$

where we regard $\mathrm{d}u$ as proportional to $\mathrm{d}z$. The assignment of the aforementioned direction cosines completes this picture, since $(\mathrm{d}x, \mathrm{d}y, \mathrm{d}u)$ is orthogonal to this surface. However, we also have the defining equation (4.1), thereby assuring that

$$\frac{\mathrm{d}x}{a} = \frac{\mathrm{d}y}{b} = \frac{\mathrm{d}u}{c}. \tag{4.4}$$

Equating these differentials with a line element $\mathrm{d}s$, we then obtain

$$\frac{\mathrm{d}x}{\mathrm{d}s} = a, \quad \frac{\mathrm{d}y}{\mathrm{d}s} = b, \quad \text{and} \quad \frac{\mathrm{d}u}{\mathrm{d}s} = c. \tag{4.5}$$

These latter equations, sometimes called the "subsidiary" equations, define the curve relating x and y, which is called a *characteristic curve*. One example of characteristic curves that is familiar to physicists are the so-called (geodesic) "world lines" that present themselves in relativity, for example. (More precisely, what we have referred to as world lines are null geodesics.) Magnetic field lines in space plasmas are another example of characteristic curves. We will encounter a practical example of these considerations later when we explore three-dimensional wave propagation. Numerical methods exist that exploit the existence of characteristic curves, but they are rarely employed apart from situations where shock fronts or material discontinuities are present. Let us now proceed to a discussion of second-order equations, which are particularly prominent in geophysical applications.

Consider the quasi-linear second-order equation

$$a\frac{\partial^2 u}{\partial x^2} + b\frac{\partial^2 u}{\partial x \partial y} + c\frac{\partial^2 u}{\partial y^2} = e$$

$$ar + bs + ct = e, \tag{4.6}$$

where it is conventional to define $r = \partial^2 u/\partial x^2$, $s = \partial^2 u/\partial x \partial y$, and $t = \partial^2 u/\partial y^2$, with p and q retaining their earlier definitions. In quasi-linear systems, a, b, and c are functions of u, p, q, x, and y but not of r, s, or t. We extend the methods we employed earlier in order to develop a suite of four equations

$$du = p\,dx + q\,dy$$
$$dp = r\,dx + s\,dy$$
$$dq = s\,dx + t\,dy$$
$$e = ar + bs + ct, \tag{4.7}$$

where the last expression is the governing partial differential equation. Let us consider the second through fourth of these four linear equations, which are necessary for the determination of r, s, and t. In particular, we note that

$$\begin{pmatrix} dx & dy & 0 \\ 0 & dx & dy \\ a & b & c \end{pmatrix} \begin{pmatrix} r \\ s \\ t \end{pmatrix} = \begin{pmatrix} dp \\ dq \\ e \end{pmatrix} \tag{4.8}$$

must be satisfied. Thus, if the determinant of the coefficient matrix above vanishes, then we have identified a condition equivalent to (4.4) above, namely,

$$
\begin{vmatrix}
\mathrm{d}x & \mathrm{d}y & 0 \\
0 & \mathrm{d}x & \mathrm{d}y \\
a & b & c
\end{vmatrix} = 0,
\tag{4.9}
$$

or

$$
c(\mathrm{d}x)^2 - b(\mathrm{d}x)(\mathrm{d}y) + a(\mathrm{d}y)^2 = 0,
\tag{4.10}
$$

The quadratic forms of hyperbolic, elliptic, and parabolic immediately follow. Simply regard the former equation as a quadratic relation between $\mathrm{d}x$ and $\mathrm{d}y$, and it becomes evident, depending on the values of a, b, and c, that one of the three quadratic forms emerges. Moreover, whenever the latter set of differential holds, no solution exists unless the other determinants resulting from this system also vanish, for example, the second through fourth equations. Thus, we obtain as a sufficient condition that

$$
\begin{vmatrix}
\mathrm{d}p & \mathrm{d}x & 0 \\
\mathrm{d}q & 0 & \mathrm{d}y \\
e & a & c
\end{vmatrix} = 0,
\tag{4.11}
$$

or

$$
e\,\mathrm{d}x\,\mathrm{d}y - a\,\mathrm{d}p\,\mathrm{d}y - c\,\mathrm{d}q\,\mathrm{d}x = 0.
\tag{4.12}
$$

Eq. (4.10) defines the characteristics while (4.12) provides the conditions that the solution must satisfy along the characteristics. We must now turn our attention to the nature of boundary conditions.

Suppose we seek the solution to a partial differential equation in u, often in higher dimension.

1. *Dirichlet conditions* apply when u is specified at each point on the boundary;

2. *Neumann conditions* apply when $(\nabla u)_n$, the normal component of the gradient of u, is specified at each point on the boundary; and

3. *Cauchy conditions* apply if u and $(\nabla u)_n$ are specified at each point of the boundary.

In general, the issue of boundary conditions can become very complicated. Morse and Feshbach (1999), Courant and Hilbert (1962), and Mathews and Walker (1970) are very helpful guides to the navigation of these questions.

We now introduce some geophysical examples that utilize these and other mathematical principles in order to derive the fundamental partial differential equations that make up much of the discipline. We begin with a simple derivation for the wave equation in one dimension to describe the behavior of a string, as a proxy for the wave propagation that is at the heart of seismology. We then derive a conservation law that is ubiquitous in geophysical fluid dynamics as well as space physics, the *continuity equation*. We will introduce an approximation known as *Fourier's law* and derive the heat flow or *diffusion equation*. We will then review how distributions of mass produce a gravitational potential, and go on to show how that integral formulation can be converted into a partial differential equation.

4.1.2 Wave Equation in One Dimension

Suppose we have a string of length L and mass M and it is held taught with a force F. For the sake of convenience, we will imagine that we can approximate the motion of the string by regarding it as a set of beads of mass m suspended at a distance a from each other so that

$$m = M\left(\frac{a}{L}\right). \tag{4.13}$$

The string is held with its end points being in the x-direction, and we wish to explore its transverse motion as described by the height h_i of the wave at mass point i. We describe this in Figure 4.1.

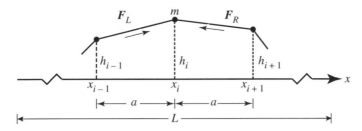

Figure 4.1. Geometry of a vibrating string.

Introducing Newton's third law, we want to describe the force that drives the oscillation of that mass point and provides the acceleration $m\ddot{h}_i$. Consider the force acting on mass point i from the segment of string extending from mass point $i - 1$.

We observe that the force on the string is of magnitude F and in the direction of the arrow shown in Figure 4.1. We need to calculate the component of the force in the vertical direction, namely, the restoring force F_L to the left of mass point i, which we can approximate for small angles by

$$F_L = \frac{h_{i-1} - h_i}{a}F, \qquad (4.14)$$

while the restoring force to the right due to mass point $i + 1$ can similarly be expressed as

$$F_R = \frac{h_{i+1} - h_i}{a}F. \qquad (4.15)$$

We equate the sum of these two forces with $m\ddot{h}_i$ and obtain

$$m\ddot{h}_i = \left(\frac{F}{a}\right)(h_{i+1} - 2h_i + h_{i-1}). \qquad (4.16)$$

We note, from Figure 4.1, that the index i refers to the longitudinal position x_i and that

$$x_{i\pm1} = x_i \pm a. \qquad (4.17)$$

It is now natural to regard the height of the wave as a function of the position x and the time t.

We can use a Taylor series to obtain a good approximation for the right-hand side of the force equation (4.16), namely,

$$h(x_{i\pm1}, t) = h(x_i, t) \pm a\frac{\partial h(x_i, t)}{\partial x} + \frac{a^2}{2}\frac{\partial^2 h(x_i, t)}{\partial x^2} + \mathcal{O}(a^3). \quad (4.18)$$

Meanwhile, we convert the acceleration term into the form $m(\partial^2 h(x_i, t)/\partial t^2)$ so that the force equation (4.16) reads, following some algebra,

$$m\frac{\partial^2 h(x_i, t)}{\partial t^2} = \left(\frac{F}{a}\right)a^2\frac{\partial^2 h(x_i, t)}{\partial x^2}. \qquad (4.19)$$

We now proceed to the limit that m and a are infinitesimally small and exploit (4.13) to obtain

$$\frac{\partial^2 h(x, t)}{\partial t^2} = \frac{FL}{M}\frac{\partial^2 h(x, t)}{\partial x^2}, \qquad (4.20)$$

where we have dispensed with the i subscript, which is no longer needed. We observe that the multiplying constant FL/M has the dimensions of a force multiplied by a distance (yielding an energy) divided by a mass; hence, it has the dimensions of a velocity squared, which we will designate by c^2, where c is the propagation velocity. Examples of the propagation velocity include the sound speed and the speed of light, depending on the circumstance. The wave equation is an example of a hyperbolic equation and is ubiquitous in geophysics and related disciplines and is generally expressed as

$$\frac{1}{c^2} \frac{\partial^2 h(\boldsymbol{x}, t)}{\partial t^2} = \nabla^2 h(\boldsymbol{x}, t), \tag{4.21}$$

where h is the scalar quantity (or a component of a vector quantity) which is undergoing wave motion. We have used the idealized behavior of a string as a proxy to describe wave motion, as it applies in seismology, geophysical fluid dynamics, electromagnetism, and other fields.

Before proceeding, we need to address two additional issues that emerge in defining the wave equation and its solution. Referring back to Newton's second law, we recall that we need to know the initial position and velocity of the object being described. Equivalently, in the wave equation, we need to know both the amplitude $h(x, 0)$ and its time derivative $h_t(x, 0)$ at initial time, which we have selected here to be $t = 0$. These so-called *initial conditions* are usually expressed as

$$h(x, 0) = a(x)$$
$$h_t(x, 0) = w(x), \tag{4.22}$$

where $a(x)$ and $w(x)$ are known functions describing the initial amplitudes and velocities. To complete this picture, we must also specify the domain over which this description applies and what happens at the boundaries, say at $x = 0$ and $x = L > 0$. Accordingly, we introduce *boundary conditions*, which are usually expressed in the form

$$h(0, t) = h_{\text{left}}(t)$$
$$h(L, t) = h_{\text{right}}(t). \tag{4.23}$$

Importantly, these boundary conditions must also be consistent with the initial conditions (4.22).

Momentarily, we will move on to higher spatial dimension and the wave equation for vectorial quantities. The wave equation also emerges for tensorial quantities, for example, stress and strain in continuum mechanics, which we will not describe here—see, for example, Newman (2012). However, there is an important concept regarding wave propagation in one dimension that we wish to introduce here. (Similar concepts emerge in higher dimension, but will not be treated in this text.) Let us now return to the wave equation (4.20), which we now express as

$$\frac{\partial^2 h(x,t)}{\partial t^2} = c^2 \frac{\partial^2 h(x,t)}{\partial x^2}. \tag{4.24}$$

It is natural to expect, from any position x, that we will see the wave amplitude proceed both to the right and to the left at a velocity of c. Hence, let us now define co-moving coordinates u and v according to

$$u = x - ct$$
$$v = x + ct. \tag{4.25}$$

It is straightforward to show that Eq. (4.24) can be written

$$\frac{\partial^2 h(u,v)}{\partial u \partial v} = 0, \tag{4.26}$$

where we have replaced the roles of x and t by u and v. A complete solution to this equation can be written

$$h(u,v) = f(u) + g(v) \tag{4.27}$$

for *any* functions f and g. Equivalently, we can write

$$h(x,t) = f(x - ct) + g(x + ct). \tag{4.28}$$

The functions f and g can now be obtained in terms of the initial condition–related functions a and w, which will be left as an exercise for the reader.

In order to visualize what this means, consider Figure 4.2 illustrating the two characteristic curves. In the left panel, segment A, we describe how wave motion proceeds at a velocity of c from an arbitrary point to the left and to the right. In the right panel, segment B, we observe how any point receives information at a velocity of c from its left and its right. These ideas are also very relevant in higher dimension, but must remain as a topic to be addressed in more advanced textbooks such as Jackson (1999).

Most geophysics graduate students, especially those focusing on space physics, have encountered Maxwell's equations (Jackson, 1999) for the electric field E and magnetic field B in

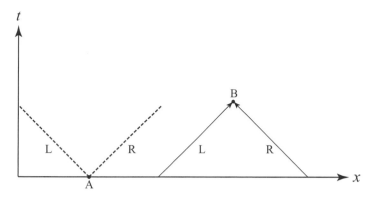

Figure 4.2. Characteristic curves.

a vacuum, where c is the speed of light employing cgs units, namely,

$$\frac{\partial \boldsymbol{E}}{\partial t} = c\, \boldsymbol{\nabla} \times \boldsymbol{B}, \qquad \boldsymbol{\nabla} \cdot \boldsymbol{E} = 0$$

$$-\frac{\partial \boldsymbol{B}}{\partial t} = c\, \boldsymbol{\nabla} \times \boldsymbol{E}, \qquad \boldsymbol{\nabla} \cdot \boldsymbol{B} = 0. \qquad (4.29)$$

By taking the time derivative of the first partial differential equation and making use of the second partial differential equation subject to the divergence conditions, we can derive immediately the wave equation for the electric field. Similarly, by taking the time derivative of the second and using the first partial differential equation, the wave equation for the magnetic field emerges. Now, we will turn our attention to a feature of flows, which is universal for conserved quantities, en route to understanding the other primary forms of partial differential equations.

4.1.3 Elements of Fluid Flow

Suppose we have a fluid that we can characterize by its mass density $\rho(\boldsymbol{x}, t)$. (We can use this same argument to describe electrical charge in place of mass, as well as energy density, and so on; literally, the flow of any conserved quantity.) In this instance, the quantity conserved is the total mass M, which can be defined by

$$M = \int_V \rho(\boldsymbol{x}, t)\, \mathrm{d}^3 x = \int_V \rho(\boldsymbol{x}, t)\, \mathrm{d}x_1\, \mathrm{d}x_2\, \mathrm{d}x_3, \qquad (4.30)$$

where V is the volume that for all times contains all of the mass. We want to understand how the mass contained in the differential volume $dx_1\,dx_2\,dx_3$ changes over a time increment dt. We illustrate this using Figure 4.3.

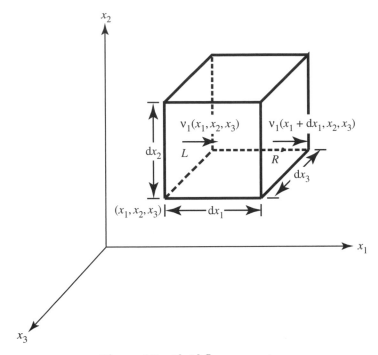

Figure 4.3.   Fluid flow geometry.

We wish to examine the change of mass in that volume and seek to approximate that change using a Taylor series, namely,

$$\rho(x_1, x_2, x_3, t + dt)\,dx_1\,dx_2\,dx_3$$
$$= \rho(x_1, x_2, x_3, t)\,dx_1\,dx_2\,dx_3$$
$$+ \frac{\partial \rho(x_1, x_2, x_3, t)}{\partial t}\,dt\,dx_1\,dx_2\,dx_3$$
$$+ dx_1\,dx_2\,dx_3\,\mathcal{O}(dt^2). \qquad (4.31)$$

From Figure 4.3, it follows that the change is due to the flux of material into and out of the cube through its six faces, two faces associated with each direction.

Let us consider the two faces associated with the x_1 direction. On the left side of the cube, designated L, a volume is swept out by the flow during the time interval dt equal to $v_1(x_1, x_2, x_3, t)\,dt\,dx_2\,dx_3$, corresponding to a mass influx of

$$\rho(x_1, x_2, x_3, t)v_1(x_1, x_2, x_3, t)\,dt\,dx_2\,dx_3.$$

Using a similar argument for the left side of the cube, designated R, the mass exiting is equal to

$$\rho(x_1 + dx_1, x_2, x_3, t)v_1(x_1 + dx_1, x_2, x_3, t)\, dt\, dx_2\, dx_3.$$

We now employ a Taylor series expansion around x_1 and obtain that the *net* mass increase from flow in the x_1 direction is

$$-\frac{\partial}{\partial x_1}[\rho(x_1, x_2, x_3, t)v_1(x_1, x_2, x_3, t)]\, dt\, dx_1\, dx_2\, dx_3.$$

Equivalent expressions emerge in the other directions, namely, x_2 and x_3. Equating the sum of these expressions with the mass change observed in the cube (4.31), and collecting and removing the terms in $dt\, dx_1\, dx_2\, dx_3$, we obtain

$$\frac{\partial \rho(\boldsymbol{x}, t)}{\partial t} = -\frac{\partial}{\partial x_1}[\rho(\boldsymbol{x}, t)v_1(\boldsymbol{x}, t)] - \frac{\partial}{\partial x_2}[\rho(\boldsymbol{x}, t)v_2(\boldsymbol{x}, t)]$$

$$- \frac{\partial}{\partial x_3}[\rho(\boldsymbol{x}, t)v_3(\boldsymbol{x}, t)], \qquad (4.32)$$

which is more commonly expressed as

$$\frac{\partial \rho(\boldsymbol{x}, t)}{\partial t} + \nabla \cdot [\rho(\boldsymbol{x}, t)\boldsymbol{v}(\boldsymbol{x}, t)] = 0. \qquad (4.33)$$

This expression is known as the *continuity equation* and is widely used to describe any conservative phenomenon where ρ is interpreted to be the density of some conserved quantity. Although often overlooked, there exists an important class of problems where mass is not conserved, notably in earth science applications involving radioactive decay (Figure 4.4). In those situations, the 0 on the right-hand side is replaced by terms that describe, for example, the decay rate of that particular isotope. Commonly, we refer to the quantity $\rho(\boldsymbol{x}, t)\boldsymbol{v}(\boldsymbol{x}, t)$ as the *mass flux*, referring to the quantity of mass that flows through a unit area in a unit of time. In the case of fluids, it is sometimes designated by the symbol $\boldsymbol{J}(\boldsymbol{x}, t)$.

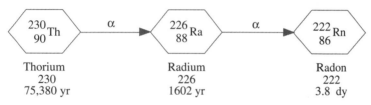

Figure 4.4. Alpha particle decay of thorium into radium and radon.

To see how this equation is conservative, consider Eq. (4.30) for the mass. Let us now integrate (4.33) over a volume V that contains all of the mass and bounded by a surface S through which no material flows. Then, we observe that

$$\int_V \frac{\partial \rho(\boldsymbol{x},t)}{\partial t}\,\mathrm{d}^3 x + \int_V \boldsymbol{\nabla} \cdot [\rho(\boldsymbol{x},t)\boldsymbol{v}(\boldsymbol{x},t)]\,\mathrm{d}^3 x = 0 = \frac{\mathrm{d}M(t)}{\mathrm{d}t}.$$
$$(4.34)$$

The term incorporating a divergence did not contribute, due to Gauss's theorem, because

$$\int_V \boldsymbol{\nabla} \cdot [\rho(\boldsymbol{x},t)\boldsymbol{v}(\boldsymbol{x},t)]\,\mathrm{d}^3 x = \int_A \rho(\boldsymbol{x},t)\boldsymbol{v}(\boldsymbol{x},t) \cdot \hat{\boldsymbol{n}}\,\mathrm{d}^2 x = 0 \quad (4.35)$$

since the flow velocity through the surface vanishes.

Assuming that the continuity equation (4.33) holds, it is useful to rewrite it in the form

$$\frac{\partial \rho(\boldsymbol{x},t)}{\partial t} + \boldsymbol{v}(\boldsymbol{x},t) \cdot \boldsymbol{\nabla}\rho(\boldsymbol{x},t) + \rho(\boldsymbol{x},t)\boldsymbol{\nabla} \cdot \boldsymbol{v}(\boldsymbol{x},t) = 0. \quad (4.36)$$

Since the flow results in particles at position $\boldsymbol{x}$ at time t moving with velocity $\boldsymbol{v}(\boldsymbol{x},t)$, it follows that the first two terms describe the convection of the material. Hence, it is useful to define a *convective derivative* designated by either $\mathrm{d}/\mathrm{d}t$ or, sometimes, $\mathrm{D}/\mathrm{D}t$ according to

$$\frac{\mathrm{d}}{\mathrm{d}t} \equiv \frac{\partial}{\partial t} + \boldsymbol{v}(\boldsymbol{x},t) \cdot \boldsymbol{\nabla}. \quad (4.37)$$

Accordingly, we can rewrite (4.33) as

$$\frac{\mathrm{d}\rho(\boldsymbol{x},t)}{\mathrm{d}t} + \rho(\boldsymbol{x},t)\boldsymbol{\nabla} \cdot \boldsymbol{v}(\boldsymbol{x},t) = 0. \quad (4.38)$$

We regard the leading term in this expression as reflecting the *total* derivative or the total change in $\rho(\boldsymbol{x},t)$ due to its explicit change in time together with the change induced by the flow in which it is embedded. With that in mind, we note that we can now write (4.37) as

$$\frac{\mathrm{d}\ln\rho(\boldsymbol{x},t)}{\mathrm{d}t} + \boldsymbol{\nabla} \cdot \boldsymbol{v}(\boldsymbol{x},t) = 0. \quad (4.39)$$

Thus, we observe that the sign of the divergence of the velocity establishes whether the density is locally contracting (positive sign) or expanding (negative sign) at an exponential rate. Finally, we incorporate the potential for radioactive decay from the specific isotope identified with ρ, with rate constant y, according to

$$\frac{\mathrm{d}\rho(\boldsymbol{x},t)}{\mathrm{d}t} + \rho(\boldsymbol{x},t)\boldsymbol{\nabla} \cdot \boldsymbol{v}(\boldsymbol{x},t) = -y\rho(\boldsymbol{x},t). \tag{4.40}$$

By comparing the divergence of the velocity with the rate constant, we can assess the relative importance of both terms, which, in principle, could be important in the Earth's interior.

Before proceeding further, it is convenient to introduce at this time another equation emerging from geophysical fluid dynamics, the *Euler force* equation. In essence, it is a statement of Newton's second law of motion for an infinitesimal volume of mass and is generally expressed as

$$\rho(\boldsymbol{x},t)\frac{\mathrm{d}\boldsymbol{v}(\boldsymbol{x},t)}{\mathrm{d}t} = \rho(\boldsymbol{x},t)\left\{\frac{\partial\boldsymbol{v}(\boldsymbol{x},t)}{\partial t} + [\boldsymbol{v}(\boldsymbol{x},t) \cdot \boldsymbol{\nabla}]\boldsymbol{v}(\boldsymbol{x},t)\right\}$$
$$= -\rho(\boldsymbol{x},t)\boldsymbol{\nabla}\Phi(\boldsymbol{x},t) - \boldsymbol{\nabla}P(\boldsymbol{x},t). \tag{4.41}$$

On the left-hand side of the equation, we observe that the acceleration involves not only the rate of change of the velocity vector at a given point but the advection of material from nearby, just as we encountered in deriving the continuity equation. On the right-hand side of the equation, we provide two of the principal forces that may apply. In the first case, we include the potential Φ whose gradient drives the acceleration, for example, under the influence of gravity in geophysics, as well as the pressure P whose gradient serves a similar role. Indeed, a balance between these two terms often emerges in more complex geophysical environments and can be related to phenomena relating to geologic overburden and to geostrophic balance in atmospheric and oceanic flows. (We will not explicitly address the issue of viscosity here, but will incorporate it into one of the assignment questions.) We shall return to the Euler force equation later when we address issues pertinent to fluid flows. Now, we turn our attention to the transport of heat.

As already noted, this conservation law is ubiquitous and we shall now use it to derive the heat equation or, as it is commonly known, the diffusion equation. Let us consider an incompressible solid or fluid with density ρ and temperature T. We wish to explore the change ΔQ in the material's internal heat (thermal energy), which varies according to

$$\Delta Q = \rho c_p \Delta T, \tag{4.42}$$

where c_p is the specific heat capacity of the medium, ΔT is the departure in temperature from some baseline, and, taken together, describe the change in the thermal energy density. Meanwhile, it is known empirically, and is often called *Fourier's law*, that the *heat flux* varies as $-\kappa \nabla T$, where κ is known as the *thermal conductivity*. This is a linear assumption good for many fluids and solids, albeit not for metals, and describes how heat flows from hotter to cooler regimes.

Taking all of these conditions into account, the continuity equation governing thermal energy becomes

$$\rho c_p \frac{\partial T}{\partial t} = \nabla \cdot (\kappa \nabla T), \qquad (4.43)$$

or, as it is often written (so long as κ remains constant),

$$\frac{\partial T}{\partial t} = D \nabla^2 T, \qquad (4.44)$$

where D is the diffusion constant defined by

$$D = \frac{\kappa}{\rho c_p}, \qquad (4.45)$$

which has the dimensions of a distance squared divided by time or, equivalently, a velocity times a distance. The diffusion equation is an example of a parabolic equation. In many materials, D quantitatively approximates the sound speed times a mean-free-path for molecules to travel. We now turn to the third common type of partial differential equation, which emerges from problems where a potential energy function, such as Φ encountered earlier, and a mass density, such as ρ, are involved.

We will now explore the diffusion and wave equations in three spatial dimensions, applying integral transform methods, to learn how to employ Fourier-based techniques and contour integration to solve complicated problems. We will then proceed to consider the nature of dispersion and diffusion.

4.2   Three-Dimensional Applications

In this section, we will present the complete solution in three dimensions of the diffusion equation, the wave equation, and the gravitational potential equations. Here, we will also develop the method of Green's functions.

4.2.1 Diffusion Equation in Three Dimensions

Let us now extend the diffusion equation (3.88) to three dimensions, namely,

$$\frac{\partial T(\boldsymbol{r}, t)}{\partial t} = D\nabla^2 T(\boldsymbol{r}, t), \tag{4.46}$$

where D is the *thermal diffusivity*, employing the initial condition

$$T(\boldsymbol{r}, 0) = T_0(\boldsymbol{r}). \tag{4.47}$$

Consider its spatial Fourier transform pair

$$F(\boldsymbol{k}, t) = \frac{1}{(2\pi)^3} \int \exp(-\mathrm{i}\boldsymbol{k} \cdot \boldsymbol{r}) T(\boldsymbol{r}, t) \, \mathrm{d}^3 r$$

$$T(\boldsymbol{r}, t) = \int \exp(\mathrm{i}\boldsymbol{k} \cdot \boldsymbol{r}) F(\boldsymbol{k}, t) \, \mathrm{d}^3 k, \tag{4.48}$$

where the doubly infinite integrals extend over three dimensions. Our initial condition results in

$$F(\boldsymbol{k}, 0) = \frac{1}{(2\pi)^3} \int \exp(-\mathrm{i}\boldsymbol{k} \cdot \boldsymbol{s}) T(\boldsymbol{s}, 0) \, \mathrm{d}^3 r, \tag{4.49}$$

which, in combination with (4.46), yields

$$\frac{\partial F(\boldsymbol{k}, t)}{\partial t} = -Dk^2 F(\boldsymbol{k}, t). \tag{4.50}$$

The latter has the immediate solution

$$F(\boldsymbol{k}, t) = \exp(-Dk^2 t) F(\boldsymbol{k}, 0). \tag{4.51}$$

Intuitively, this statement indicates that each wavenumber associated with the temperature distribution undergoes an exponential decay at a rate proportional to k^2 and the diffusion coefficient D. We can now introduce this result in the second of our transform pair equations to obtain

$$T(\boldsymbol{r}, t) = \int \exp(\mathrm{i}\boldsymbol{k} \cdot \boldsymbol{r}) F(\boldsymbol{k}, 0) \exp(-Dk^2 t) \, \mathrm{d}^3 k$$

$$= \frac{1}{(2\pi)^3} \iint \exp[\mathrm{i}\boldsymbol{k} \cdot (\boldsymbol{r} - \boldsymbol{s})] T_0(\boldsymbol{s}) \exp(-Dk^2 t) \, \mathrm{d}^3 k \, \mathrm{d}^3 s. \tag{4.52}$$

We take note that

$$k^2 = k_1^2 + k_2^2 + k_3^2$$

$$\boldsymbol{k} \cdot (\boldsymbol{r} - \boldsymbol{s}) = k_1(r_1 - s_1) + k_2(r_2 - s_2) + k_3(r_3 - s_3). \tag{4.53}$$

It is evident that the integration over d^3k can be expressed as three independent integrals over dk_1, dk_2, and dk_3. Thus, we observe that we can define

$$
\begin{aligned}
G(\boldsymbol{r}, t; \boldsymbol{s}) = \frac{1}{2\pi} &\int \exp[ik_1(r_1 - s_1)] \exp(-Dk_1^2 t)\, dk_1 \\
\times \frac{1}{2\pi} &\int \exp[ik_2(r_2 - s_2)] \exp(-Dk_2^2 t)\, dk_2 \\
\times \frac{1}{2\pi} &\int \exp[ik_3(r_3 - s_3)] \exp(-Dk_3^2 t)\, dk_3.
\end{aligned}
\tag{4.54}
$$

We recognize that we are taking three independent Fourier transforms of a Gaussian, as we had in addressing (3.88), and the result becomes

$$
G(\boldsymbol{r}, t; \boldsymbol{s}) = \frac{1}{(4\pi Dt)^{3/2}} \exp\left[-\frac{|\boldsymbol{r} - \boldsymbol{s}|^2}{4Dt}\right].
\tag{4.55}
$$

Further we note that

$$
\lim_{t \to 0} G(\boldsymbol{r}, t; \boldsymbol{s}) = \delta(\boldsymbol{r} - \boldsymbol{s}) = \delta(r_1 - s_1)\delta(r_2 - s_2)\delta(r_3 - s_3).
\tag{4.56}
$$

We introduce this Green's function into (4.52), whereupon we observe

$$
\begin{aligned}
T(\boldsymbol{r}, t) &= \int G(\boldsymbol{r}, t; \boldsymbol{s}) T_0(\boldsymbol{s})\, d^3 s \\
&= \int \frac{1}{(4\pi Dt)^{3/2}} \exp\left[-\frac{|\boldsymbol{r} - \boldsymbol{s}|^2}{4Dt}\right] T_0(\boldsymbol{s})\, d^3 s.
\end{aligned}
\tag{4.57}
$$

We now turn our attention to the wave equation in three dimensions.

4.2.2   Wave Equation in Three Dimensions

We consider the wave equation in an infinite region, but remain cognizant of the relationship between the source position $\boldsymbol{x}'$ and the observer position $\boldsymbol{x}$, as well as the time associated with the source, t', and the observer, t. Accordingly, we wish to solve

$$
\nabla^2 \psi - \frac{1}{c^2}\frac{\partial^2 \psi}{\partial t^2} = \delta(\boldsymbol{x} - \boldsymbol{x}')\delta(t - t').
\tag{4.58}
$$

We have been explicit in the manner in which we have expressed this in order to emphasize that the solution depends upon both $\boldsymbol{x} - \boldsymbol{x}'$ and $t - t'$. We must now employ Fourier transform pairs in space and in time. Accordingly, we write

$$\psi(\mathbf{x}, t) = \int \frac{d^3k}{(2\pi)^3} \int \frac{d\omega}{2\pi} \Psi(\mathbf{k}, \omega) \exp[i(\mathbf{k} \cdot \mathbf{x} - \omega t)]$$

$$\Psi(\mathbf{k}, \omega) = \int d^3x \int dt \, \psi(\mathbf{x}, t) \exp[-i(\mathbf{k} \cdot \mathbf{x} - \omega t)]. \qquad (4.59)$$

We now employ the Fourier transform of our differential equation to obtain

$$\left(-k^2 + \frac{\omega^2}{c^2}\right) \Psi(\mathbf{k}, \omega) = 1. \qquad (4.60)$$

The "1" on the right-hand side is the outcome of integrating the two δ functions over all space and time.

The result of this calculation is that

$$\Psi(\mathbf{k}, \omega) = \frac{c^2}{\omega^2 - k^2 c^2}. \qquad (4.61)$$

Consequently,

$$\psi(\mathbf{x}, t) = c^2 \int \frac{d^3k}{(2\pi)^3} \int \frac{d\omega}{2\pi} \frac{\exp[i(\mathbf{k} \cdot \mathbf{x} - \omega t)]}{\omega^2 - k^2 c^2}. \qquad (4.62)$$

We now introduce polar coordinates in $\mathbf{k}$-space oriented along the $\mathbf{x}$ vector. The integration around the azimuthal ϕ coordinate yields a factor of 2π, which cancels immediately one of those present here. The integration along the azimuthal coordinate (equivalent to $\cos\theta$ or μ) yields two terms. These can be combined into one by extending the range of integration of k, which is originally from 0 to ∞, to the range from $-\infty$ to ∞. The outcome of these manipulations, where we use r in place of $|\mathbf{x}|$, is

$$\psi(\mathbf{x}, t) = \frac{1}{(2\pi)^3} \frac{c^2}{ir} \int_{-\infty}^{\infty} k \, dk \int_{-\infty}^{\infty} d\omega \frac{\exp[i(kr - \omega t)]}{\omega^2 - k^2 c^2}, \qquad (4.63)$$

where the integration variable k is no longer regarded as a polar or radial variable but one that varies from minus to plus infinity.

We now perform the integral over ω and note that there are two poles on the ω axis, one situated at $-|kc|$ and one situated at $|kc|$, as shown in Figure 4.5.

Note the positions of the two poles present in this problem. This uncertainty is an outcome of our failure to specify boundary conditions. We have a choice of going under, over, or through each of these poles, and in each case we would pick up some

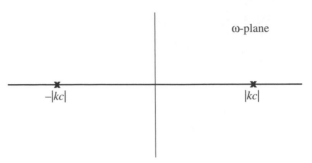

Figure 4.5.   Contour in the ω-plane employed in calculating the Green's function.

measure of

$$\frac{\exp[i(kr \pm \omega t)]}{r}.$$

These possible additions from these poles are simply homogeneous solutions to Helmholtz's equation (2.131) that we observed earlier. Evidently, our focus must be upon the influence of a disturbance localized, say, at $t = 0$ and at $\boldsymbol{x} = 0$, just as in the diffusion equation. We seek to introduce *causality* into our prescription: We assume that $\psi = 0$ for $t < 0$. This presumes that no effect is experienced, that is, nothing happens, before the disturbance. [Some sources, such as Jackson (1999), refer to two solutions emerging, calling them "retarded" and "advanced," respectively. The retarded solution corresponds to the disturbance occurring before it is observed, while the advanced solution is noncausal and is generally ignored.] So, we make our contour go over both poles and we complete the contour by an upper semicircle when $t < 0$, thereby enclosing no poles. But, when $t > 0$, we complete the contour by a lower semicircle, yielding

$$\psi(\boldsymbol{x}, t) = \frac{1}{(2\pi)^3} \frac{c^2}{ir} \int_{-\infty}^{\infty} k\, dk \exp(ikr) \int_{-\infty}^{\infty} d\omega \frac{\exp(-i\omega t)}{\omega^2 - k^2 c^2}$$

$$= \frac{1}{(2\pi)^3} \frac{c}{2ir} \int_{-\infty}^{\infty} dk \exp(ikr)$$

$$\times \int_{-\infty}^{\infty} d\omega \exp(-i\omega t) \left[\frac{1}{\omega - kc} - \frac{1}{\omega + kc} \right]$$

$$= -\frac{c}{8\pi^2 r} \int_{-\infty}^{\infty} dk \exp(ikr)[\exp(-ikct) - \exp(ikct)]$$

$$= -\frac{c}{4\pi r}[\delta(r - ct) - \delta(r + ct)]. \qquad (4.64)$$

Since both $r = |\boldsymbol{x}|$ and t are positive, the second δ function never contributes and the emergent Green's function is just

$$G(\boldsymbol{x} - \boldsymbol{x}', t - t')$$

$$= \begin{cases} 0 & \text{if } t < t', \\ -\dfrac{c}{4\pi|\boldsymbol{x} - \boldsymbol{x}'|}\delta[|\boldsymbol{x} - \boldsymbol{x}'| - c(t - t')] & \text{if } t > t'. \end{cases}$$

$$(4.65)$$

There is a simple physical interpretation for this result. The solution for ψ falls off as $1/r$, as one expects from the potential equation, but the source of the disturbance that is felt corresponds to the (earlier) time associated with the signal being propagated over the intervening distance.

In order to see how this Green's function applies, consider the solution to

$$\nabla^2\psi(\boldsymbol{x}, t) - \frac{1}{c^2}\frac{\partial^2\psi(\boldsymbol{x}, t)}{\partial t^2} = f(\boldsymbol{x}, t). \qquad (4.66)$$

Introducing the Green's function, we then obtain

$$\psi(\boldsymbol{x}, t) = -\frac{c}{4\pi}\int \mathrm{d}^3x'\,\mathrm{d}t'\,f(\boldsymbol{x}', t')\frac{\delta[|\boldsymbol{x} - \boldsymbol{x}'| - c(t - t')]}{|\boldsymbol{x} - \boldsymbol{x}'|}$$

$$= -\frac{1}{4\pi}\int \mathrm{d}^3x'\frac{f(\boldsymbol{x}', t - |\boldsymbol{x} - \boldsymbol{x}'|/c)}{|\boldsymbol{x} - \boldsymbol{x}'|}. \qquad (4.67)$$

This, again, is the *retarded potential* since the term in the integral corresponds to a source occurring at an earlier time, according to the distance over which the signal had to be propagated. In applications to the electrodynamics of a moving charge, this result gives rise to the *Lienard–Wiechert potential*, which lies beyond the scope of this text, although some aspects of the problem such as *relativistic aberration* and the *Poynting–Robertson effect* (Harwit, 2006; Stacey and Davis, 2008) emerge in considering the evolution of dust grains in the solar system. The reader is referred to Mathews and Walker (1970) or Jackson (1999) for a treatment of this problem.

We now return to the question emerging from gravitational or, equivalently, magnetostatic potentials, which we will approach using spherical harmonics and derive the associated Green's function.

4.2.3   Gravitational Potential and Green's Function Methods

Recall Poisson's equation (2.125):

$$\nabla^2 \Phi(\boldsymbol{x}) = 4\pi G \rho(\boldsymbol{x}). \qquad (2.125).$$

Poisson's equation is an example of an elliptic equation. Suppose that we now assume that we can express ρ and Φ, respectively, by

$$\rho(\boldsymbol{x}) = \rho(r,\theta,\phi) = \sum_{\ell=0}^{\infty} \sum_{m=-\ell}^{\ell} \rho_{\ell,m}(r) Y_{\ell,m}(\theta,\phi)$$

$$\Phi(\boldsymbol{x}) = \Phi(r,\theta,\phi) = \sum_{\ell=0}^{\infty} \sum_{m=-\ell}^{\ell} \Phi_{\ell,m}(r) Y_{\ell,m}(\theta,\phi). \qquad (4.68)$$

These relations are readily invertible in terms of obtaining $\rho_{\ell,m}$ and $\Phi_{\ell,m}$ due to the orthonormality of the $Y_{\ell,m}$. We recall the identity

$$\nabla^2 Y_{\ell,m} = -\frac{\ell(\ell+1)}{r^2} Y_{\ell,m}. \qquad (4.69)$$

Taking the Laplacian of the potential, it immediately follows that

$$\frac{1}{r^2}\frac{d}{dr}\left[r^2 \frac{d\Phi_{\ell,m}(r)}{dr}\right] - \frac{\ell(\ell+1)}{r^2}\Phi_{\ell,m}(r) = 4\pi G \rho_{\ell,m}(r). \qquad (4.70)$$

This is a second-order linear inhomogeneous differential equation, which we have treated at length in chapter 3. We immediately observe that the homogeneous solutions to this are r^{ℓ} and $r^{-(\ell+1)}$. In order to obtain a solution to the inhomogeneous problem, we employ the method of variation of constants presented earlier.

We will assume that the solution for $\Phi_{\ell,m}(r)$ can be expressed

$$\Phi_{\ell,m}(r) = C_{\ell,m}(r)r^{\ell} + D_{\ell,m}(r)r^{-(\ell+1)}. \qquad (4.71)$$

As before, we will assume that, where primes $'$ designate derivatives with respect to r,

$$C'_{\ell,m}(r)r^{\ell} + D'_{\ell,m}(r)r^{-(\ell+1)} = 0, \qquad (4.72)$$

as in the usual Euler–Lagrange methodology. As before, we will obtain two equations in two unknowns, the first being the preceding expression and the second being

$$C'_{\ell,m}(r)[\ell r^{\ell-1}] + D'_{\ell,m}(r)[-(\ell+1)r^{-(\ell+2)}] = 4\pi G \rho_{\ell,m}(r). \qquad (4.73)$$

Solving between these two equations, we obtain

$$C'_{\ell,m}(r)(2\ell + 1)r^{\ell-1} = 4\pi G\rho_{\ell,m}(r) \tag{4.74}$$

and

$$D'_{\ell,m}(r)[-(2\ell + 1)r^{-(\ell+2)}] = 4\pi G\rho_{\ell,m}(r). \tag{4.75}$$

We can now combine these results to obtain the complete solution, namely,

$$\Phi_{\ell,m}(r) = -\int_{r'=r}^{\infty} \frac{4\pi G\rho_{\ell,m}(r')}{(2\ell + 1)} \frac{r^{\ell}}{r'^{(\ell+1)}} r'^2 \, dr'$$
$$-\int_{r'=0}^{r} \frac{4\pi G\rho_{\ell,m}(r')}{(2\ell + 1)} \frac{r'^{(\ell)}}{r^{\ell+1}} r'^2 \, dr'. \tag{4.76}$$

We can render this expression more compact by writing

$$\Phi_{\ell,m}(r) = -\int_0^{\infty} G\rho_{\ell,m}(r')G_\ell(r,r')r'^2 \, dr', \tag{4.77}$$

where the Green's function is defined by

$$G_\ell(r,r') = \frac{4\pi}{2\ell + 1}
\begin{cases}
\dfrac{r'^{(\ell)}}{r^{(\ell+1)}} & \text{if } r' < r, \\[2ex]
\dfrac{r^{(\ell)}}{r'^{(\ell+1)}} & \text{if } r' > r.
\end{cases} \tag{4.78}$$

We observe that both parts of the solution remain continuous at $r' = r$ but have a discontinuity in derivative of $2\ell + 1$, which is offset by the preceding fractional term. A useful way to think of this formulation, as expressed, for example, by Jackson (1999), is that we select the smaller of the two radii r and r' inside the integral to appear in the numerator while the larger of the two appears in the denominator. Both of these terms preserve the form of the eigenfunctions r^{ℓ} and $r^{-(\ell+1)}$.

A key ingredient in this discussion relates to the transition in the solution resulting from the application of these formulae to the calculation of the gravitational potential for the Earth and other celestial bodies, as measured outside their surface. In particular, suppose that the density $\rho(r)$, and hence $\rho_{\ell,m}(r)$, vanishes for $r > R$, where R is effectively the radius of that planet. The formulae above guarantee that both the potential and its gradient—since we are describing a three-dimensional problem—vary continuously across the interface between the interior of that planet and its exterior. Moreover, we note outside the planet that the $C_{\ell,m}$ and $D_{\ell,m}$ coefficients are now constant since there is no source term $\rho_{\ell,m}$ present to cause them to change; this outcome can immediately be confirmed by inspecting (4.72), (4.74), and (4.75). Hence, we need only perform a quadrature of (4.75) to obtain $D_{\ell,m}(R)$, which are the values we employ for those constants outside that planet in representing its gravitational potential, and note that $C_{\ell,m}(R)$ is necessarily zero in order for the solution outside the planet to remain well behaved as $r \rightarrow \infty$. This observation has a remarkable outcome—since $D_{\ell,m}(R)$ can be established via a quadrature of the density distribution inside the planet, we have no information available through these coefficients alone as to the structure of that planet's interior. Indeed, the planet could be a hollow but nonuniform shell and present the same gravitational potential. The only way to overcome this indeterminacy is through the construction of models for that planet's interior or through other measurable quantities, such as the seismic signature, heat flow, and magnetic field variation.

4.3 Diffusion, Dispersion, Perturbation Methods, and Nonlinearity

We now proceed to consider the nature of diffusion and dispersion as well as some elements of perturbation theory in application to simple nonlinear partial differential equations resulting from sound waves. We will derive some fundamental results emerging from solitary wave and soliton propagation, as well as introduce the concepts of self-similarity, scaling, and the Kolmogorov theory of turbulence.

4.3.1 Diffusion and Dispersion

Earlier, we treated the homogeneous wave equation (4.60). We observed that the frequency ω was algebraically related to the wavenumber k, that is,

$$\omega^2 = k^2 c^2. \tag{4.79}$$

This kind of expression is generally referred to as a *dispersion relation*. Typically, it is much more complicated, especially if some form of dissipation is present or there are more than two underlying partial differential equations. For example, when we introduced the one-dimensional wave equation (4.24) for a vibrating string, we could have introduced a viscous damping term associated with the string vibrating in some medium. In that instance, we would then have

$$\frac{\partial^2 h(x,t)}{\partial t^2} + v\frac{\partial h(x,t)}{\partial t} = c^2\frac{\partial^2 h(x,t)}{\partial x^2}, \tag{4.80}$$

where v defines the rate at which the velocity is reduced due to friction. This expression is known as the *telegraph equation* (Strang, 1986), which has its origins in the theory underlying transmission lines and coaxial cables. Fourier transforming this expression, we observe that its dispersion relation is given by

$$\omega^2 + iv\omega = k^2 c^2, \tag{4.81}$$

which clearly implies that ω has a real and an imaginary part.

Electromagnetic theory also provides a compelling basis for its derivation. If we take the first of the Maxwell equations (4.29) and add a current density term $4\pi J$ to the left-hand side—as in problem (11)—and assume that the current satisfies *Ohm's law*

$$J = \sigma E, \tag{4.82}$$

where σ is the electrical conductivity, then it is easy to show that

$$\omega^2 + 4\pi i\sigma\omega = k^2 c^2, \tag{4.83}$$

where k is now the vectorial wavenumber. (A more precise treatment of the displacement current is required, however, to prevent the wave speed from exceeding that of light (Jackson, 1999).) It follows immediately that v in (4.81) for viscous-damped wave motion is equivalent to $4\pi\sigma$ in the electromagnetic problem.

We can now solve this problem for ω and observe, under normal circumstances where $v^2 < 4k^2 c^2$, that

$$\omega = -\frac{iv}{2} \pm \frac{1}{2}\sqrt{4k^2 c^2 - v^2}. \tag{4.84}$$

Accordingly, we observe that ω has a real and an imaginary part. The imaginary part corresponds to an exponential decay in time, namely, $\exp(-vt/2)$, which is generally referred to as "diffusion," although that is in part a misnomer, while it is also appropriate to refer to it as "dissipation." However, the real part corresponds to the $\pm kc$ dominated behavior in the wave equation in the absence of dissipation but there are some important differences.

In particular, the real part of ω, which we will designate ω_r, is no longer proportional to k—we will focus on the scalar properties in the dispersion relation, but note that there are problems where the vector properties can be important. The real and imaginary parts of the dispersion relation can be related using the so-called *Kramers–Kronig relation* (Arfken and Weber, 2005). Wave propagation depends upon both the "phase" preserved between the spatial and temporal terms, as well as the degree of coherence preserved between the spatial and temporal terms. For example, we can define a *phase velocity* v_φ by trying to preserve the phasing of the x and t contributions to

$$kx - \omega_r(k)t = k[x - v_\varphi(k)t], \tag{4.85}$$

from which we identify

$$v_\varphi(k) = \frac{\omega_r(k)}{k} = \frac{\sqrt{4k^2c^2 - v^2}}{2k}. \tag{4.86}$$

The phase velocity, on the other hand, is not directly connected to transport properties and can be illusory.

Before proceeding further, there is a beautiful physical example of the illusory nature of phase velocity. Imagine standing on a long beach immediately behind the water line, and imagine a wave approaching the beach from some distance. However, the wave peak is along a line that is not exactly parallel to the water line on the beach. If we focus on the speed of the wave peak striking the beach along the water line, it can be extremely large, yet the speed of the water itself is moderate. This is an illustration of the nature of the phase relationship that appears.

Feynman et al. (1989) present an elegant example that we adapt here. Consider two interfering waves with equal amplitude having the form

$$\psi(x,t) = \exp[i(k_1 x - \omega_1 t)] + \exp[i(k_2 x - \omega_2 t)]. \tag{4.87}$$

This can be rewritten as

$$\psi(x,t) = \exp\left\{\frac{i}{2}[(k_1 + k_2)x - (\omega_1 + \omega_2)t]\right\}$$
$$\times \left[\exp\left\{\frac{i}{2}[(k_1 - k_2)x - (\omega_1 - \omega_2)t]\right\} \right.$$
$$\left. + \exp\left\{-\frac{i}{2}[(k_1 - k_2)x - (\omega_1 - \omega_2)t]\right\} \right].$$
$$\tag{4.88}$$

We recognize the first term as representing, in effect, the mean of the two waves, while the added terms describe the modulation that occurs due to the wavenumber and frequency differences. It follows that the speed of the wave is still essentially ω/k, but the modulation now introduces a new "velocity," which is

$$v_g \approx \frac{\omega_1 - \omega_2}{k_1 - k_2}. \tag{4.89}$$

This is the speed of propagation of the modulation, which, in turn, describes how coherent the two waves remain. The envelope associated with the modulation describes how well the two waves work cooperatively.

We have just seen how the phase velocity does not describe how contributions to the wave from different wavenumbers preserve the coherence of a "group" of nearby wavenumbers. In order to illustrate this, we now regard ω as a function of k and, when we write $\omega(k + dk)$, it is a function of $k + dk$:

$$[k + dk]x - [\omega(k + dk)]t = k[x - v_\varphi(k)t] + dk[x - v_g(k)t], \tag{4.90}$$

where we have employed a Taylor series expansion and where the *group velocity* $v_g(k)$ is defined by

$$v_g(k) = \frac{d\omega_r(k)}{dk} = \frac{2c^2 k}{\sqrt{4k^2 c^2 - v^2}}. \tag{4.91}$$

It follows from (4.90) that the group velocity describes how rapidly x and t change relative to each other in preserving the identity of some feature. Finally, in this case, we also observe that

$$v_\varphi(k)v_g(k) = c^2, \tag{4.92}$$

for the dispersion relation (4.84). While this latter result is a direct outcome of the particular dispersion relation at hand, it is not uncommon to find phase velocities that are greater than the speed of light (or seismic velocities in solid Earth applications), while the group velocities are smaller than those physically limiting velocities. The group velocity, which relates to how different wavenumbers can maintain coherence, describes the transport of energy between otherwise interfering parts of a wave. We will return later to issues of diffusion and to dispersion. Meanwhile, we wish to explore atmospheric sound waves (Houghton, 2002) to better understand their properties and to present a vehicle for exploring the characteristics of flow.

Before departing from this section, let us return briefly to the issue of diffusion and the issue of self-similarity, and the heat equation (3.88)

$$\frac{\partial T(x,t)}{\partial t} = D\frac{\partial^2 T(x,t)}{\partial x^2}$$

and its Green's function solution (3.95)

$$T(x,t) = \frac{1}{\sqrt{4\pi Dt}} \int_{-\infty}^{\infty} dy\, T_0(y) \exp\left[-\frac{(x-y)^2}{4Dt}\right].$$

From strictly dimensional considerations, it follows from Eq. (3.88) that the distribution grows over a characteristic time scale τ to a characteristic length scale ℓ according to

$$\ell \approx \sqrt{D\tau}. \tag{4.93}$$

The Green's function solution (3.95) demonstrates this in a powerful way inasmuch as the Green's function broadens the original temperature distribution at time t, due to the appearance of the Gaussian, by an amount of order $\sqrt{Dt}$. Indeed, if the original temperature distribution has a characteristic size $\mathcal{L}$, it follows that after a time of order $\mathcal{L}^2/D$, the Gaussian broadening will take over. Thereafter, *all* solutions will appear to have a Gaussian shape and thereby become self-similar. The term *self-similarity* emerged from considerations wherein the solution would evolve to one where its "shape" would remain unchanged in time. For example, consider the density of ink particles as a drop of ink released into a large vessel diffuses through it. The variation in color—which relates to the shape—assumes a uniform character while the intensity or amplitude of the coloration dissipates over time. In the (linear) diffusion problem, therefore, we would expect the functional dependence in space to evolve according to the scaled variable $x/\sqrt{Dt}$.

Self-similarity can manifest in nonlinear problems as well (Newman, 1983b, 1984). We observe that Eq. (3.88) has two conservation laws. The first conserved quantity is the total energy, which is proportional to $\int_{-\infty}^{\infty} T(x,t)\,dx$. The second conserved quantity is the first moment of the temperature distribution, which is $\int_{-\infty}^{\infty} xT(x,t)\,dx / \int_{-\infty}^{\infty} T(x,t)\,dx$. Without loss of generality, we will move the coordinate origin in our calculation to

coincide with the first moment of the distribution, thereby eliminating an unimportant length scale from the problem. Barenblatt and Zel'Dovich (1972) developed the general theory for
self-similar solutions, especially those emerging in cases with
conservations law, which they called "self-similar solutions of
type I." [Type II solutions correspond to problems without conservation laws where the self-similar solution is the outcome of a
nonlinear eigenvalue problem. Barenblatt and Zel'Dovich (1972)
also showed that traveling-wave problems could be converted
into self-similar ones.] Barenblatt (1996) elaborated on this, but
did not establish a methodology for establishing that all solutions to equations possessing self-similar scalings converge to
the self-similar solution. Newman (1983b, 1984) demonstrated
that many diffusion problems with conservation laws ultimately
converge to self-similarity, thereby establishing them as *global
attractors* for the problem. Newman (1980, 1983a,b) showed that
certain nonlinear diffusion problems possessing traveling-wave
solutions had this long-time evolutionary character.

Let us again consider Eq. (3.88) with the first moment of the
temperature distribution situated at the origin; it must necessarily spread out, as we have already observed via the Green's function Eq. (3.95). Diffusion, as in the ink drop example, causes the
distribution to lose all memory of its initial conditions—apart
from its total energy and its first moment—and must necessarily evolve to a universal shape dictated by the underlying partial
differential equation. Let us proceed to show this.

Apart from the $x/\sqrt{Dt}$ spatial scaling, we expect that the
amplitude of the temperature distribution scales as $1/\sqrt{Dt}$. The
latter follows since the temperature distribution will spread out
over a length scale $\sqrt{Dt}$ but, as the integrated thermal energy
is conserved, must therefore drop in amplitude by that amount.
This can also be seen in the Green's function solution Eq. (3.95).
Without loss of generality, we introduce a scale-free "spatial"
variable

$$y = \frac{x}{\sqrt{Dt}}, \tag{4.94}$$

and a scale-free "time" variable

$$\tau = \ln(t/t_0), \tag{4.95}$$

where $t_0 > 0$ is some characteristic time. Then, we will introduce a scale-free temperature variable $G(y, \tau)$ according to

$$T(x, t) \equiv \frac{G(y, \tau)}{\sqrt{Dt}}. \qquad (4.96)$$

This transformation is completely general since it preserves both spatial and temporal variability in the new temperature variable. However, we must now convert our partial differential equation in the independent variables x and t to an equation in the new independent variables y and τ. To do this for any dependent variable $g(x, t) \rightarrow g(y, \tau)$, we employ the differential relationship

$$dg = \frac{\partial g}{\partial y} \, dy + \frac{\partial g}{\partial \tau} \, d\tau, \qquad (4.97)$$

from which we obtain

$$\frac{\partial g}{\partial x} = \frac{1}{\sqrt{Dt}} \frac{\partial g}{\partial y}$$

$$\frac{\partial^2 g}{\partial x^2} = \frac{1}{Dt} \frac{\partial^2 g}{\partial y^2} \quad \text{and}$$

$$\frac{\partial g}{\partial t} = -\frac{y}{2t} \frac{\partial g}{\partial y} + \frac{1}{t} \frac{\partial g}{\partial \tau}. \qquad (4.98)$$

Equating $g(y, \tau)$ with $G(y, \tau)/\sqrt{Dt}$, we get

$$\frac{\partial G(y, \tau)}{\partial \tau} = \frac{\partial^2 G(y, \tau)}{\partial y^2} + \frac{y}{2} \frac{\partial G(y, \tau)}{\partial y} + \frac{G(y, \tau)}{2}$$

$$= \frac{\partial}{\partial y} \left[\frac{\partial G(y, \tau)}{\partial y} + \frac{y}{2} G(y, \tau) \right]. \qquad (4.99)$$

If the temperature distribution converges to self-similarity, then $G(y, \tau)$ will only depend upon y. Thus, the left-hand side would vanish, rendering the term inside the square brackets on the far right-hand side, namely,

$$\frac{\partial G(y, \tau)}{\partial y} + \frac{y}{2} G(y, \tau),$$

a constant. Since the total energy remains finite, G must vanish as $y \rightarrow \pm\infty$, thereby making the bracketed quantity vanish as well. Accordingly, the self-similar solution $G_{ss}(y)$ satisfies

$$G_{ss}(y) = c \exp(-y^2/4), \qquad (4.100)$$

where the proportionality constant c is selected to match the energy conservation requirement. It is noteworthy that this preserves the exponential character obtained in the Green's function solution (3.95). Making use of Eq. (3.4), we observe that

$$\int_\infty^\infty G_{ss}(y)\,dy = c\sqrt{4\pi} = \sqrt{Dt_0}\int_{-\infty}^\infty T(x,t_0)\,dx, \qquad (4.101)$$

where the latter follows from our relationship Eq. (4.96) connecting $T(x,t)$ with $G(y,t)$.

Given the thermodynamic basis of the heat equation, it is natural to expect that some entropy-like argument might be invoked to establish convergence to self-similarity—see Newman (1983b, 1984). Consider writing Eq. (4.99) as

$$\frac{\partial G(y,\tau)}{\partial\tau} = \frac{\partial}{\partial y}\left\{G(y,\tau)\frac{\partial}{\partial y}\left[\ln G(y,\tau) + \frac{y^2}{4}\right]\right\}. \qquad (4.102)$$

This expression exposes another method for isolating the self-similar solution: As before, either $G(y,\tau)$ is 0 or $\ln G(y,\tau) + y^2/4$ is a constant. Then, let us define a *Lyapunov functional* $H(\tau)$ by

$$H(\tau) = -\int_{-\infty}^\infty G(y,\tau)\left[\ln G(y,\tau) + \frac{y^2}{4}\right]dy, \qquad (4.103)$$

and calculate its derivative with respect to τ, that is,

$$\begin{aligned}
\frac{\partial H(\tau)}{\partial\tau} &= -\int_{-\infty}^\infty \left[\ln G(y,\tau) + 1 + \frac{y^2}{4}\right]\frac{\partial G(y,\tau)}{\partial\tau}\,dy \\
&= \int_{-\infty}^\infty G(y,\tau)\left\{\frac{\partial}{\partial y}\left[\ln G(y,\tau) + \frac{y^2}{4}\right]\right\}^2 dy \\
&\geqslant 0
\end{aligned} \qquad (4.104)$$

after integrating by parts (Newman, 1984). This establishes that $H(\tau)$ increases everywhere monotonically until either $G(y,\tau)$ vanishes or $\ln G(y,\tau) + y^2/4$ becomes constant with respect to y for $G(y,\tau) \neq 0$. As before, we observe that this is *exactly* the condition in Eq. (4.102) for self-similarity to be achieved. To interpret the Lyapunov functional, we observe "if we regard $G(y,\tau)$ as a probability" that the energy (density) is situated at "position" y at "time" τ whose classical entropy is just $-G(y,\tau)\ln G(y,\tau)$. The $G(y,\tau)y^2/4$ term is then a measure of how widely the energy is distributed in a spatial sense. See Newman (1983b, 1984) for more details regarding this methodology and its interpretation, as well as Newman (1980, 1983a) regarding demonstration of global attractors of a diffusive traveling-wave flow. We now proceed to explore a nondissipative form of traveling-wave behavior, that associated with the propagation of sound waves, and present a brief introduction to perturbation theory.

4.3.2   Sound Waves and Perturbation Theory

Let us now introduce the Euler force equation in one dimension for the velocity $v(x, t)$ as

$$\frac{dv}{dt} + \frac{1}{\rho}\frac{\partial P}{\partial x} = 0, \tag{4.105}$$

where ρ and P are the density and pressure, respectively. This is equivalent to (4.41) in one spatial dimension x with no gravitational or other potential present. We now present the continuity equation (4.33) in advective form in one dimension:

$$\frac{d\rho}{dt} + \rho\frac{\partial v}{\partial x} = 0. \tag{4.106}$$

After some algebraic manipulation, the first law of thermodynamics for an adiabatic gas with adiabatic index y is

$$\frac{d}{dt}\left[\frac{P}{\rho^y}\right] = 0. \tag{4.107}$$

The latter equation has the effect of introducing a third characteristic into the problem emerging from the first two, namely,

$$\frac{dx}{dt} = v. \tag{4.108}$$

The physical interpretation of this is obvious: This characteristic describes how entropy remains unchanged when describing an individual atom or molecule, the *sine qua non* of adiabatic processes. This result also allows us to rewrite the continuity equation as

$$\frac{1}{y}\frac{d\ln P}{dt} + \frac{\partial v}{\partial x} = 0. \tag{4.109}$$

We must now explore the behavior of the other equations.

Let us now assume that the flow can be described by a mean or unperturbed (constant) value, designated using a 0 subscript plus perturbations designated by a subscript of 1. (We could employ a small expansion parameter ε as we did earlier in our treatment of the pendulum problem. The situation here is sufficiently simple that there is no added benefit to that.) Hence, we will write

$$v(x,t) = v_0 + v_1(x,t)$$
$$P(x,t) = P_0 + P_1(x,t)$$
$$\rho(x,t) = \rho_0 + \rho_1(x,t). \tag{4.110}$$

We will assume that the perturbation quantities in the density and the pressure are smaller than the respective unperturbed values. Keeping only linear contributions from the perturbations, we obtain that

$$\left(\frac{\partial}{\partial t} + v_0 \frac{\partial}{\partial x}\right) v_1 + \frac{1}{\rho_0} \frac{\partial P_1}{\partial x} = 0$$

$$\left(\frac{\partial}{\partial t} + v_0 \frac{\partial}{\partial x}\right) P_1 + \gamma P_0 \frac{\partial v_1}{\partial x} = 0. \tag{4.111}$$

We can eliminate v_1 between these two equations to get

$$\left(\frac{\partial}{\partial t} + v_0 \frac{\partial}{\partial x}\right)^2 P_1 = \left(\frac{\gamma P_0}{\rho_0}\right) \frac{\partial^2 P_1}{\partial x^2}, \tag{4.112}$$

We now recognize this as the wave equation in a spatial coordinate that is moving with speed v_0, that is, the unperturbed flow, and where the sound-speed v_s is given by

$$v_s = \sqrt{\frac{\gamma P_0}{\rho_0}}, \tag{4.113}$$

consistent with the conventional result from thermodynamics that

$$v_s^2 = \frac{dP(\rho)}{d\rho}. \tag{4.114}$$

The analysis of this equation yields two additional characteristics, namely,

$$\frac{dx}{dt} = v_0 \pm v_s, \tag{4.115}$$

which identifies that sound propagation proceeds in either direction at a speed v_s relative to the atom or molecule in question, while the third characteristic that we identified proceeds at a speed v_0.

It is easy to obtain using Fourier analysis the dispersion relation that applies here, and we obtain after some algebra that

$$(\omega - kv_0)[(\omega - kv_0)^2 - k^2 v_s^2] = 0. \tag{4.116}$$

We see a direct relationship here between the dispersion relation, and its associated phase and group velocities, and the characteristics we had obtained earlier. In geophysical applications, many other partial differential equations emerge, with many being multidimensional, and with many containing other variable quantities. The methods of perturbation theory, nonetheless, remain the same: We identify the unperturbed solution, linearize the equations present, and identify the relevant dispersion relations. Upon doing that, the roles of dispersion and diffusion become clear.

Before departing from the issue of sound waves, it is useful to introduce the notion of the *decibel*, which essentially is the logarithm multiplied by 10 of the amplitude of the sound wave and typically is designated using the symbol *dB*. There are a variety of physical quantities encountered where the logarithm of the quantity is a more natural descriptor; in the case of sound, the human ear, for example, perceives the relative amplitude of different sounds in logarithmic fashion. Suppose there are three sound waves, with the second having an amplitude 10 times the first, and the third having an amplitude 10 times the second or 100 times the first. Using the dB scale, the second sound wave would be considered to be 10 dB greater than the first ($10 \times \log_{10} 10$) while the third sound wave would be considered to be 20 dB greater ($10 \times \log_{10} 100$). Another illustration of the use of logarithms familiar to geophysicists is the *Richter magnitude scale* (Stacey and Davis, 2008). Crudely speaking, an earthquake with magnitude 5.0 has a shaking amplitude 10 times that of a magnitude 4.0 event—as identified by the motion of a seismometer's arm—while the relative increase in the energy liberated is 31.6. Logarithmic scales are endemic in nature and are employed in applications determining the pH of a fluid (for acidity and alkalinity), the brightness of stars (stellar magnitude), octaves in music (which employ $\log_2$ of the frequency), and many other examples. Now, we proceed to explore some nonlinear problems that have yielded important insights into real-world systems.

4.3.3 Burgers's Equation and Solitary Waves

One of the leading sources of nonlinearity in geophysical problems is the advective or last term in

$$\frac{d\boldsymbol{v}}{dt} = \frac{\partial \boldsymbol{v}}{\partial t} + (\boldsymbol{v} \cdot \boldsymbol{\nabla})\boldsymbol{v} \tag{4.117}$$

that appears in the Euler force equation (4.41). In particular, we have seen earlier how the characteristic for this equation is $dx/dt = v$, which can readily result in shock-like behavior. For example, consider in one dimension a flow where $\partial v/\partial x < 0$ and $v > 0$; this is the situation that appears on a sloping beach where inbound water coming from a distance does so at greater speed than water closer to shore and wave-breaking occurs as the faster, more distant flow overtakes the slower, closer-at-hand flow. As we have observed earlier in an exercise, the *vorticity* $\omega \equiv \nabla \times v$ also preserves some important flow properties and, as we shall see, resides at the heart of turbulent behavior. Once again, the advective term in the force equation plays a central role.

In order to physically arrest the wave-breaking and other shock-like behaviors, it was natural to consider introducing a diffusion or viscosity term, namely,

$$\frac{\partial v(x,t)}{\partial t} + v(x,t) \cdot \frac{\partial v(x,t)}{\partial x} = D\frac{\partial^2 v(x,t)}{\partial x^2}, \qquad (4.118)$$

which we shall consider in an unbounded medium subject to the initial condition

$$v(x,0) = v_0(x). \qquad (4.119)$$

This equation was first developed by Burgers (1948) as a mathematical model for turbulence, and it now bears his name. Whitham (1974) provides a detailed treatment of Burgers's equation, and Newman (2000) describes its role in turbulent (inverse) cascades. The analysis method we are about to present was developed by Cole et al. (1951) and Hopf (1950) and often is referred to as the Cole–Hopf solution.

We now introduce the Cole–Hopf transformation

$$v(x,t) = -2D\frac{\theta_x(x,t)}{\theta(x,t)}. \qquad (4.120)$$

After some algebra, we observe that this transforms Burgers's equation into the linear diffusion equation

$$\frac{\partial \theta(x,t)}{\partial t} = D\frac{\partial^2 \theta(x,t)}{\partial x^2}. \qquad (4.121)$$

We observe that this is exactly the partial differential equation (3.88) that we solved earlier for temperature diffusion. Importantly, given the role of diffusion, the distribution $\theta(x,t)$ is smoothed out over time, thereby assuring that $v(x, t)$ does not develop any kind of discontinuity.

In order to illustrate this, we plot in Figure 4.6 the solution to Burgers's equation (4.118) for a step function–like discontinuity as its original condition with the initial velocity on the left-hand side $u_0 = 1$.

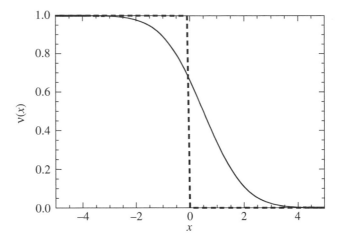

Figure 4.6. Solution to Burgers's equation
with $D = 1$ and $v_0(x) = H(-x)$ at $t = 1$.

From dimensional analysis arguments, we expect that the "thickness" of the shock front to be of order $D/u_0 = 1$.

While this result is not readily extendable to higher spatial dimension, it suggests nevertheless that certain kinds of nonlinearity can lend themselves to the formation of coherent structures. These are generally called *solitary waves* and are localized phenomena that seemingly maintain their structure. This will be the focus of our next section.

4.3.4   Korteweg–de Vries Equation and Solitons

A classic example of this is the Korteweg–de Vries equation, which has its origins, in the absence of nonlinearity, in the dimensionless equation

$$v_t + v_x + v_{xxx} = 0, \tag{4.122}$$

which we observe to be a linearized version of Burgers's equation, assuming that the unperturbed solution there is $v_0(x, t) = 1$ and has a dispersion relation of

$$\omega = k - k^3. \tag{4.123}$$

It is evident that small wavenumbers, that is, long wavelengths, travel at a relatively uniform speed and that the phase and group velocities slow with increasing wavenumber. This, then, offers some interesting possibilities as to how nonlinearities could help regulate the overall dispersion of localized structures.

The solitary wave was evidently first observed by J. Scott Russell riding alongside the Edinburgh-Glasgow canal in 1834 past a stopped boat (Drazin and Johnson, 1989). Russell observed the formation of

> a large solitary elevation, a rounded, smooth and well-defined heap of water, which continued its course along the channel apparently without change of form or diminution of speed. (pp. 7–8)

Russell performed laboratory experiments with shallow-water waves, with the formation of long (compared to their depth) solitary waves that travelled at a well-defined speed closely related to their amplitude, the effect of nonlinearity. Moreover, he observed that individual solitary waves of this type could pass through each other unscathed, suffering only a modest deceleration or acceleration in the process.

Before proceeding further with the mathematical modeling of shallow water waves as well as their atmospheric equivalent, gravity waves, a few comments relating to dimensional relations are in order. The gravitational acceleration g, which has the units of meters/second^2, and the depth h of the water channel, which has the units of meters, are all important. We note that $\sqrt{gh}$ has the units of a velocity in meters/second and is a good approximation to the speed of wave propagation. For ocean water, with abyssal depths $\approx 5000\,\mathrm{m}$, given that $g = 9.8\,\mathrm{m/s^2}$, speeds in excess of $200\,\mathrm{m/s}$ or $800\,\mathrm{km/hr}$ are expected. This is the basis for *tsunamis* (Stacey and Davis, 2008). In atmospheric flows, the scale height of the atmosphere is the appropriate measure of atmospheric thickness (Houghton, 2002), essentially the vertical distance over which an effectively isothermal atmosphere sustains a pressure drop of $1/e$, and is approximately $10\,\mathrm{km}$. This quantity is given by $H = kT/mg$, where k is the Boltzmann constant and m is the mean molecular weight, and $\sqrt{gH}$ turns out to be comparable to the thermal velocity of atmospheric molecules. Finally, the quantity $\sqrt{g/h}$ has the dimensionality of a frequency and corresponds to the oscillations known as gravity waves. In atmospheric applications, this is known as the *Brunt–Väisälä frequency* (Houghton, 2002) and the corresponding time scale is 30 s. While our focus in the next example pertains to the applicable mathematics in these dispersive environments, the quantitative magnitudes of these phenomena are well worth remembering.

This stimulated many researchers, including Boussinesq and Lord Rayleigh, who also observed a seemingly universal shape for the disturbance, with Boussinesq (1877) performing further analysis. However, the literature generally refers to Korteweg and de Vries in 1895 as having identified the governing partial differential equation, which is now generally written using the standard form (Drazin and Johnson, 1989; Ablowitz and Segur, 1981)

$$v_t - 6v v_x + v_{xxx} = 0. \tag{4.124}$$

This equation enjoys a number of similarities with Burgers's equation, replacing a diffusive term, namely, v_{xx}, by a dispersive one, v_{xxx}, and it was hoped that ideas emergent from the Cole–Hopf transformation would be helpful here. However, many surprises were left to be discovered. Drazin and Johnson (1989) and Ablowitz and Segur (1981) provide comprehensive historical as well as mathematical treatments of this problem.

The analysis of this problem begins by seeking to identify a *traveling wave* solution, namely,

$$v(x,t) = f(x - ct), \tag{4.125}$$

where f is a negative quantity and describes the functional form of the solution and c is the speed of propagation. We will now obtain the ordinary differential equation defining f—the assumption of a traveling wave has reduced the problem from that of a partial differential equation—and the speed at which it travels. In general, c can be a (nonlinear) eigenvalue. In this circumstance, we will observe that it presents an intriguing control over the shape of the solution, including its amplitude and the width of the pulse that forms. Applying (4.125) to (4.124), we obtain

$$-c\frac{\mathrm{d}f}{\mathrm{d}x} - 6f\frac{\mathrm{d}f}{\mathrm{d}x} + \frac{\mathrm{d}^3 f}{\mathrm{d}x^3} = 0, \tag{4.126}$$

which can be rewritten

$$\frac{d}{dx}\left[-cf - 3f^2 + \frac{d^2f}{dx^2}\right] = 0. \qquad (4.127)$$

It is evident that the quantity in brackets is a constant, a first integral for the problem. However, we anticipate at great distance from the localized disturbance that f and, hence, f'', must vanish, leaving us with

$$-cf - 3f^2 + \frac{d^2f}{dx^2} = 0. \qquad (4.128)$$

We now multiply by df/dx and group terms to obtain yet another integral, namely,

$$\frac{df}{dx}\left[-cf - 3f^2 + \frac{d^2f}{dx^2}\right] = \frac{d}{dx}\left[-c\frac{f^2}{2} - f^3 + \frac{1}{2}\left(\frac{df}{dx}\right)^2\right]. \qquad (4.129)$$

We recognize that the quantity in square brackets is a constant that also vanishes in order for the disturbance to disappear at great distance. Therefore, we are left with

$$\left(\frac{df}{dx}\right)^2 = f^2(c + 2f). \qquad (4.130)$$

We have now reduced the solution to a quadrature, namely,

$$x - ct - x_0 = -\int \frac{df}{f\sqrt{2f + c}} = -\int \frac{d\sqrt{2f + c}}{f}, \qquad (4.131)$$

where we have introduced an additive constant to the left-hand side and taken explicit note that the argument of f was $x - ct$ in describing the traveling wave.

We now make the algebraic substitution

$$g^2 = \frac{2f + c}{c}, \qquad (4.132)$$

and obtain

$$x - ct - x_0 = -\frac{2}{\sqrt{c}}\int \frac{dg}{g^2 - 1}. \qquad (4.133)$$

Finally, we make the substitution

$$g = \coth u, \qquad (4.134)$$

where the solution is reduced to

$$x - ct - x_0 = 2u/\sqrt{c}, \qquad (4.135)$$

finally yielding

$$f(x - ct) = -\frac{c}{2}\operatorname{sech}^2\left[\frac{\sqrt{c}}{2}(x - ct - x_0)\right]. \tag{4.136}$$

We immediately observe an intriguing feature of this solution: Its amplitude is proportional to c while its width varies as $1/\sqrt{c}$. Further, the negative value of this solution reflects our choice of sign for the advective term in (4.124), which we selected to conform with convention. Employing a positive sign instead would make this solution positive. As an illustration, selecting $c = 1$, we present the solution for $-f(x)$ in Figure 4.7.

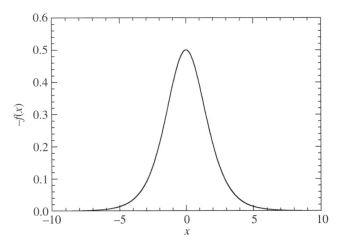

Figure 4.7.   Solution to Korteweg–de Vries equation with $c = 1$.

The solution to the Korteweg–de Vries equation (4.124) presents many surprises. Not only does it preserve its shape when it travels at a constant speed, that is, has a traveling wave solution, but the interaction or "collision" of two such pulses moving at different speeds results in the two pulses passing through each other and then emerging, restored to their original, independent shapes with their original speeds but with a possible "phase shift" in the locations of their centers. Hirota (1971) discovered an exact solution to the Korteweg–de Vries equation including multiple solitons, which we illustrate in Figure 4.8 using a set of parameters that we selected to make clear the nature of the interaction.

In this figure, we display as before $-v(x, t)$ at uniform intervals of time displaced vertically so that each of the five presentations are clearly identifiable, beginning with the solution at

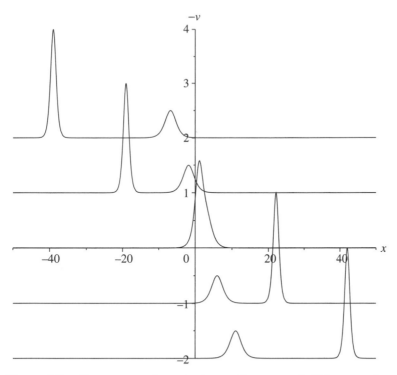

Figure 4.8. Hirota two-pulse solution to Korteweg–de Vries equation.

the earliest time showing nearest the top and then proceeding downwards. In each of the curves, we see two pulses moving to the right, with the higher amplitude one moving with 4 times the speed of the of the smaller amplitude pulse, which is 1/4 of its height. The larger amplitude pulse overtakes the smaller one and the two have effectively merged in the middle display. They then separate again and distance themselves from each other in the lower pair of curves.

As an added means of showing how two independent pulses interact, the contour plot in Figure 4.9 visually presents the Hirota solution employing contours. The two pulses are evident, but, when they interact, their relative "phase" seems to change with the faster pulse accelerating and the slower pulse decelerating before resuming their constant velocity trajectories. Accordingly, it was universally recognized that solutions to the Korteweg–de Vries equation and a limited number of other partial differential equations presented a very special class of solitary waves that became known as *solitons*.

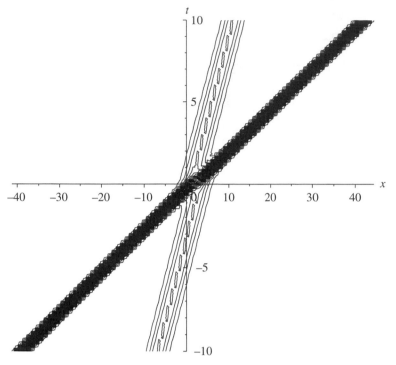

Figure 4.9.   Hirota two-pulse solution to Korteweg–de Vries
equation demonstrating phase lag.

It was observed that the Korteweg–de Vries equation had an
infinite number of conservation laws, not simply the two we
identified in solving for the traveling wave solution. The scope of
these discoveries extends far beyond this text but is comprehen-
sively described in Drazin and Johnson (1989) and Ablowitz and
Segur (1981). In particular, it was found that the solution to the
Korteweg–de Vries equation satisfied a special form of nonlinear
Schrödinger equation whose infinite set of eigenvalues consti-
tuted the conserved quantities. Accordingly, since algebraically
similar problems appear in scattering theory, these discoveries
produced the so-called *inverse scattering transform*, or *IST*.

There is a very rich literature on this topic, but solitons remain
a rather elusive and very special form of solitary wave. A fundamental feature of solitons is the existence of an infinite number of conservation laws that determine their behaviour, but,
from a practical point of view, the preservation of the shape of
individual solitons as they pass through each other and regain
their former velocities is the dominant feature. In geophysical

applications, idealized soliton behavior is very rare, if indeed it exists. However, the geophysical literature provides many illustrations of their possible presence in a variety of environments such as compacting partially molten magmatic systems (Scott and Stevenson, 1984, 1986; Scott et al., 1986; McKenzie, 1987; Fowler, 2011).

The fundamental lesson here, however, is that many phenomena can produce solitary waves that preserve to some degree their identity through balancing nonlinear and dispersive effects. The discovery of solitons appeared around the same time as the identification of fractal geometry with many natural phenomena and the idea that simple dynamical models could engender chaos. The combination of these different kinds of behavior has given rise to the concept of *complexity*, the notion that high degree-of-freedom systems can demonstrate, in some instances, remarkable sensitivity to initial conditions—the hallmark of chaos in low degree-of-freedom systems. Complex systems also yield geometrical aspects that defy traditional modes of thinking, yet can also present collective modes of behavior in others that underscore a cooperation between different scales. We will turn our attention now to the latter class of problems, in application to *turbulence* and scaling phenomena, which we saw earlier were the signature aspect of fractal geometry.

4.3.5 Self-Similarity, Scaling, and Kolmogorov Turbulence

Among the most enduring nonlinear phenomena that characterize modern geophysics is turbulence in fluids. Turbulence is a nonlinear mechanism that transports energy, typically from the largest relevant size scales down to the smallest relevant ones, possibly extending over many orders of magnitude. (The behavior in many geophysical phenomena is more complicated owing to the introduction of a variety of intervening size scales.) This phenomenon is very much at the heart of the picture that motivated the development of the Lorenz model. When convection is initiated in a vessel, it typically creates eddies or vortices on the largest scale. As the flow becomes more vigorous, the vortices break down into smaller structures, down to the molecular level, where viscous or thermal dissipation takes place. Texts such as those by Tennekes and Lumley (1972) and Newman (2012) provide a review of some of the underlying phenomenology, while Batchelor (1953) provides a more technically detailed investigation. Monin et al. (2007) have developed a comprehensive and up-to-date treatment of the subject. Turbulent flow is an inherently random process where probabilistic or statistical considerations are central. Moreover, vortices occupy a pivotal role. Remarkably, although possessing an intrinsic directionality, their overall distribution must become isotropic in a statistical and global sense during the cascade that we are about to describe. We can visualize vortices being formed at the scale of the vessel in which the fluid is present, for example, the atmosphere. This largest size scale is referred to as the *production subrange*. From those largest size scales, energy is transferred systematically to vortices of smaller size scale, a process sometimes referred to as a *direct cascade*. Owing to the transfer of energy through a sequence of successively smaller vortices, this dynamic process extending often through orders of magnitude of scale is referred to as the *inertial subrange*. Finally, as the smallest vortices ultimately yield to individual molecules where heat can be liberated, we refer to the *dissipation range*. Not all cascades, however, are direct. In some planetary atmospheres, they can be indirect, wherein small eddies merge together to form larger ones. Newman (2000) showed how this can happen via Burgers's equation.

Kolmogoroff developed some remarkable insights into turbulent cascades and how they can give rise to power-law spectra based on a simple energy-transport argument. His "statistical theory" assumed that we have an ensemble of eddies or vortices, each having a random orientation so that the overall fluid behavior is isotropic—hence, his theory is referred to as one for isotropic turbulence. In particular, suppose we have an eddy of size scale λ whose rotational speed is characteristically v_λ. Then, the time scale for one rotation of an eddy is of order λ/v_λ. Hence, the dissipation rate ϵ_λ at that scale λ—with the energy being delivered to smaller size scales—is given approximately as

$$\epsilon_\lambda \approx \frac{v_\lambda^3}{\lambda}. \tag{4.137}$$

Let us now assume that the dissipation rate in the inertial subrange is scale free, that is, $\epsilon_\lambda = \epsilon$, where ϵ is a constant that is independent of scale λ. We can invert the latter relationships to obtain

$$v_\lambda \approx (\epsilon\lambda)^{1/3}, \tag{4.138}$$

an approximation widely known as Obukhov's law. We have assumed scale invariance is present in the underlying equations and is often assured in observations of fluids. We relate the wavenumber k to the length scale λ according to $k \approx 2\pi/\lambda$, whereupon it follows that

$$v_k \approx \left(\frac{\epsilon}{k}\right)^{1/3}. \tag{4.139}$$

The power spectrum $S(k)$, or power spectral density in N dimensions, is an estimate of the quantity of energy present at a given wavenumber, and satisfies

$$S(k) \approx \frac{v_k^2}{k^N} \approx \epsilon^{2/3} k^{-N-2/3}, \tag{4.140}$$

which is generally known as the *Kolmogorov–Obukhov law*. Many complex flows in three dimensions have this power-law index of $-11/3$. Other flows, characterized, for example, by an angular momentum or vorticity cascade, will satisfy different power laws.

While this brings us to the end of our chapter on partial differential equations, we have also attempted to incorporate some ideas emergent from complexity theory (Sornette, 2006). Unlike the topic of chaos in ordinary differential equations, complexity is somewhat elusive in attempting to define it. Typically, it manifests in some sensitivity to initial conditions, but generally involves a large number of degrees of freedom. Some problems that manifest complexity have a continuum underpinning, but others do not, and the topic of *self-organized criticality* has developed widespread interest—see, for example, the books by Bak (1996), Jensen (1998), and Sornette (2006).

4.4 Exercises

1. Consider the equation encountered in fluid mechanics for momentum conservation, namely,

$$\frac{\partial v}{\partial t} + v\frac{\partial v}{\partial x} = 0.$$

Show that the characteristic equation for this equation is

$$\frac{\mathrm{d}x}{\mathrm{d}t} = v(x,t).$$

Explain the physical significance of the characteristic.

2. Consider the potential equation in two spatial dimensions

$$\frac{\partial^2 u}{\partial x^2} + \frac{\partial^2 u}{\partial y^2} = 0.$$

Show that the characteristics for this reveal that it is an elliptic equation.

3. Consider the wave equation in one spatial dimension where c is the speed of light,

$$c^2 \frac{\partial^2 u}{\partial x^2} = \frac{\partial^2 u}{\partial t^2}.$$

Show that this is a hyperbolic equation.

4. The thermal diffusion equation in one spatial dimension is

$$\frac{\partial T}{\partial t} = D \frac{\partial^2 T}{\partial x^2}$$

but has only a single second-order term, that is, the second-derivative term we identified before as r. Consider a rotation in the x-t plane and show, accordingly, that the condition connecting a, b, and c in the quasi-linear second-order equation that preserves this equation having parabolic character is

$$ac - b^2 = 0.$$

5. Consider the heat flow equation

$$\frac{\partial T(x,t)}{\partial t} = D \frac{\partial^2 T(x,t)}{\partial x^2},$$

where $T(x,t)$ is the temperature distribution in x at time t, D is the diffusion coefficient, and the initial conditions are

$$T(x,0) = T_0(x),$$

valid for $-\infty < x < \infty$.

(a) Show that $G(x,t)$, the Green's function for this problem, can be expressed

$$G(x,t;y) = \frac{\exp[-(x-y)^2/4Dt]}{\sqrt{4\pi Dt}},$$

by showing (i) that it satisfies the differential equation above, and (ii) as $t \to 0$ that it becomes a δ function in $(x-y)$. [Hint: As $t \to 0$, show that it goes to zero when $x \neq y$ and that its integral over all x is unity.]

(b) Show that

$$T(x,t) = \int_{-\infty}^{\infty} G(x,t;y)T_0(y)\,dy$$

satisfies the original differential equation (you may use the result from part a of this problem) as well as the initial conditions.

(c) Suppose that the initial condition is

$$T_0(x) = \sin(x).$$

Show that

$$T(x,t) = \exp(-Dt)\sin(x)$$

is the complete solution for this problem. [Hint: Think! You do not need to use contour integration to show this.]

6. Recall the equation for the vorticity of the flow

$$\boldsymbol{w}(\boldsymbol{x},t) = \nabla \times \boldsymbol{v}(\boldsymbol{x},t).$$

Since the cross product of two vectors is orthogonal to both of the original vectors, we might be tempted to expect that quantities such as

$$\boldsymbol{v}(\boldsymbol{x},t) \cdot \boldsymbol{w}(\boldsymbol{x},t) = 0$$

would vanish. Suppose that we have a flow defined by

$$\boldsymbol{v}(\boldsymbol{x},t) = \boldsymbol{\Omega} \times \boldsymbol{x} + (\hat{\boldsymbol{\Omega}} \cdot \boldsymbol{x})\boldsymbol{\Omega},$$

where $\boldsymbol{\Omega}$ is an arbitrary constant vector and $\hat{\boldsymbol{\Omega}}$ is a unit vector in that direction. Calculate the divergence of this velocity field. Moreover, show that $\boldsymbol{v}(\boldsymbol{x},t) \cdot \boldsymbol{w}(\boldsymbol{x},t) = 0$ is not true by calculating the true value its right-hand side.

7. Suppose that the density of a fluid undergoing flow is a constant, neither changing in time nor in position. Using the continuity equation, show that

$$\nabla \cdot \boldsymbol{v}(\boldsymbol{x},t) = 0$$

for "incompressible flow."

8. In the Euler force equation, we encounter the term

$$\{v(x,t) \cdot \nabla\}v(x,t).$$

Using the Levi-Civita symbol, show that

$$(v \cdot \nabla)v = \nabla(\tfrac{1}{2}v^2) - v \times \omega$$

using the vorticity defined earlier.

9. Suppose that we are dealing with an incompressible fluid that is subject to viscous forces, but gravity and pressure play no important role. The flow velocity $v(x,t)$ satisfies the equation

$$\frac{\partial v(x,t)}{\partial t} + \{v(x,t) \cdot \nabla\}v(x,t) = \nu\nabla^2 v(x,t),$$

where ν is the diffusion rate for velocity, and is commonly called the *kinematic viscosity*. Take the curl of of the previous equation and show that

$$\frac{\partial \omega(x,t)}{\partial t} + \{v(x,t) \cdot \nabla\}\omega(x,t)$$
$$= \{\omega(x,t) \cdot \nabla\}v(x,t) + \nu\nabla^2\omega(x,t).$$

The first term on the right-hand side of the equation is referred to as a "vortex-stretching" term in three dimensions and is fundamental to turbulence theory; unless it provides a significant transfer of energy among different length scales, turbulent behavior cannot happen.

10. Suppose we multiply the former equation by $\omega(x,t)$ and integrate over all space. Show that the enstrophy

$$\mathcal{E} \equiv \frac{1}{2}\int_V |\omega|^2 \, d^3x$$

must necessarily decrease over time, so long as the vortex-stretching term in $\omega \cdot \nabla v$ does not intervene, which it cannot (why?) for two-dimensional flows. This is the essence of the Taylor–Proudman theorem, which also has a magnetic-field analogue for strong fields known as the *Cowling theorem*. What are the implications of this result to geophysical turbulence?

11. We introduced Maxwell's equations (4.29) describing a vacuum. Taking the partial derivative with respect to time of the first two equations, prove that

$$\frac{1}{c^2}\frac{\partial^2 E}{\partial t^2} = \nabla^2 E$$

and

$$\frac{1}{c^2}\frac{\partial^2 B}{\partial t^2} = \nabla^2 B.$$

Suppose we now modify these equations to include the presence of charge density ρ_e and its current density J, again using cgs units, namely,

$$\frac{\partial E}{\partial t} + 4\pi J = c\nabla \times B, \qquad \nabla \cdot E = 4\pi \rho_e$$

$$-\frac{\partial B}{\partial t} = c\nabla \times E, \qquad \nabla \cdot B = 0.$$

Our discussion of the continuity equation (4.33) and the conservation of mass can also be applied to the conservation of electrical charge. We recognize that J describes the flux of charge through a surface per unit time, and is analogous to the mass flux ρv that we encountered earlier. Taking the divergence of the first of these equations, show that the continuity equation applied to electrical charge emerges

$$\frac{\partial \rho_e}{\partial t} + \nabla \cdot J = 0.$$

12. In astrophysics, the large-scale expansion of the universe appears to follow *Hubble's law*, which can be written

$$v = H_0 x$$

where x is the position, say of a galaxy, relative to some benchmark, v is the observed velocity, and H_0 is a constant with the dimensions of reciprocal time and is often approximated as 75 (km/s)/MPc (kilometers per second per megaparsec, a unit designed to simplify the calculation of redshift from a separation distance). (Note that 1 MPc $= 3.0857 \times 10^{22}$ m.) Suppose that this expression holds in the continuity equation as given in (4.37). Show that this implies that the density of the universe is decreasing at an exponential pace everywhere. The Hubble constant H_0 has a dimensionality of $[T]^{-1}$; calculate the corresponding time and compare it with the age of the universe (approximately 13.8×10^9 yr).

13. Consider the sound wave problem introduced earlier. However, introduce a small expansion parameter ε, and, as in the pendulum problem treated in chapter 2, expand all variables in terms of ε. For example, we would use

$$v(x,t) = v_0(x,t) + \varepsilon v_1(x,t) + \varepsilon^2 v_2(x,t) + \cdots$$

where $v_0(x,t)$ remains a constant, that is, the unperturbed flow velocity.

(a) Show that the ε^1 contribution to the solution is the same as that discussed in chapter 2.

(b) What equations must be satisfied by the ε^2 terms? Discuss your result.

14. Consider Burgers's equation

$$\frac{\partial v}{\partial t} + v\frac{\partial v}{\partial x} = \nu\frac{\partial^2 v}{\partial x^2}.$$

Show that it is a partial differential equation of mixed type having both parabolic and hyperbolic aspects.

15. Consider the diffusion equation in three dimensions with initial conditions

$$T_0(s) = A\exp[-Bs^2],$$

where A and B are positive constants. What is the solution for the temperature distribution for $t > 0$?

16. Return to the derivation we have performed using Cartesian coordinates for the solution to the diffusion equation and, in particular, its Green's function. Repeat the calculation using spherical coordinates and an appropriate extension of contour integration methods to accommodate the more complicated integrand that appears. [Hint: Our treatment of the wave equation using polar coordinates could be helpful here.]

17. Suppose you have a disturbance moving at a uniform velocity so that you can write

$$\boldsymbol{x}' = \boldsymbol{x}_0 + \boldsymbol{v}_0 t';$$

in particular, you will presume that

$$f(\boldsymbol{x}', t') = A\delta[\boldsymbol{x}'].$$

Solve for the potential associated with this moving disturbance. [Hint: Draw a picture showing where the disturbance had to be and when in order to influence the observer now.]

18. Suppose that Burgers's equation (4.118) is satisfied by $v(x, t)$ with initial conditions

$$v(x, 0) = \begin{cases} u & \text{if } x < 0, \\ 0 & \text{if } x > 0, \end{cases}$$

where $u > 0$ is a constant velocity, mimicking a shock wave. Show that the Cole–Hopf transformation–related function $\theta(x, t)$ has a corresponding initial value of

$$\theta(x, 0) = \begin{cases} \exp(-ux/2D) & \text{if } x < 0, \\ 1 & \text{if } x > 0. \end{cases}$$

19. The Korteweg–de Vries equation (4.124) has an infinite number of conservation laws.

   (a) Show that

   $$\int_{-\infty}^{\infty} v(x, t)\, dx$$

   is conserved by taking its time derivative and integrating by parts, as necessary.

   (b) Do the same for

   $$\int_{-\infty}^{\infty} v^2(x, t)\, dx.$$

   (c) Do the same for

   $$\int_{-\infty}^{\infty} \frac{v^3(x, t)}{3} + \left[\frac{\partial v(x, t)}{\partial x}\right]^2 dx.$$

We will not consider the generalization to this problem but simply note that there exist generating functions for such quantities (Drazin and Johnson, 1989; Ablowitz and Segur, 1981).

20. In our derivation of the Kolmogorov–Obukhov law, we exploited the use of dimensionless scaling arguments. These can provide very important insights into the scaling properties for linear and nonlinear equations.

   (a) For the usual one-dimensional diffusion equation

   $$\frac{\partial T}{\partial t} = D\frac{\partial^2 T}{\partial x^2},$$

   provide an argument showing that the relevant length scale ℓ goes as $\sqrt{D\tau}$, where τ is the relevant distance scale. Explain the significance of that, for example, to the diffusion of heat and the cooling time for the Earth. Assuming that the thermal distribution is spatially confined, or is at least not infinite, show that

   $$I_0 = \int_{-\infty}^{\infty} T(x,t)\,\mathrm{d}x$$

   is conserved. Show also that the "first moment" μ of the temperature distribution defined by I_1/I_0, where I_1 is given by

   $$I_1 = \int_{-\infty}^{\infty} xT(x,t)\,\mathrm{d}x,$$

   is also preserved. Finally, show that the "second moment" σ^2 of the temperature distribution defined by I_2/I_0, where I_2 is given by

   $$I_2 = \int_{-\infty}^{\infty} (x-\mu)^2 T(x,t)\,\mathrm{d}x,$$

   which is a measure of the distance scale ℓ squared, varies linearly in time by evaluating $\mathrm{d}I_2/\mathrm{d}t$. Show how this confirms your earlier result obtained by strictly dimensionless analysis relating distance and time scales.

   (b) Consider a special case of the dimensionless form of the porous medium equation encountered in hydrology where ρ is the density, namely,

   $$\frac{\partial \rho}{\partial t} = \frac{\partial^2 \rho^2}{\partial x^2}.$$

   First, show that $\int_{-\infty}^{\infty} \rho(x,t)\,\mathrm{d}x$ is conserved, implying that the amplitude of the density varies crudely as ℓ.

Accordingly, show that

$$\ell \propto \tau^{1/3},$$

unlike the linear problem with the 3 in the power law a consequence of the extra power of ρ in the definition of the diffusion term (unlike the usual Fick's law or Fourier's law, the flux of material is also proportional to the density).

CHAPTER FIVE

Probability, Statistics, and Computational Methods

Probability theory has had a profound role in the history of mathematics, but is also of great importance in the natural sciences and engineering inasmuch as many problem areas contain an element of randomness. The concept of fractals, especially coming about in stochastic processes, has fundamentally transformed much of the geosciences. Hamming (1991) presents a conceptual introduction to the subject that is very readable, and other books even with a semipopular focus have entered this arena. Feller (1968) remains a classic two-volume reference in the field, and is commonly employed at the graduate level. The utilization of probabilistic methods in computer simulation has become increasingly important, and sources such as Whitney (1990) provide helpful introductions to that subject. We now survey some fundamental concepts and results emerging from probability theory that are relevant to a broad range of problems, including those identified in geophysical laboratory and field observations as well as computer simulations. Finally, while many natural phenomena are describable via simple probabilistic distributions, such as the binomial, Poisson, and normal distributions, other problems give rise to so-called *fat-tailed distributions*, and we will offer some insight into such problems. Freedman et al. (2007) provides an excellent introduction to statistics and a survey of the field, including core concepts and many real-world examples.

While the analytic methods we have described allow us to explore the linear regime for many geophysical problems, non-linearity presents challenges that often can only be addressed through a careful and insightful use of the computer. Many years ago, such problems were approached by adapting the calculus via finite difference methods to specially designed computer languages such as FORTRAN, which stands for "FORmula TRANslation." Computer languages proliferated over the years with the emergence of many high-level languages, including today's C, C++, and Python. Moreover, we have seen the emergence of special-purpose programs such as Matlab, Mathematica, and Maple that allow many kinds of computation to be performed using limited programming skills and packaged routines; Maple and Mathematica introduced programming capabilities by incorporating the ability to produce procedures using a simple version of C. In addition, Mathematica and Maple offer "algebra engines" that facilitate the performance of intensive algebraic calculations, and all three offer the ability to present results in a graphical fashion using a relatively simple and user-friendly interface. Unfortunately, these special-purpose programs often provide little if any insight as to how they obtain their results; the algorithms employed are rarely specified, leaving the user with substantial uncertainties as to the accuracy, let alone the reliability, of such results.

Our purpose here is to survey some of the methods that have proven to be so helpful in application to geophysical problems, but also clarify some of their limitations. Our intent is to help the reader select the methods that are best suited for a particular problem, and offer some insight into their limitations. The *sine qua non* of scientific computation is that it parallels laboratory investigations in terms of reproducibility, repeatability, and reliability. Furthermore, we want to keep our discussion free from specific concerns emerging from the type of computer used, the language employed, or the package utilized. The computer is an important ally in our investigation, especially of nonlinear problems, but due diligence is required in obtaining results in which we can maintain confidence.

5.1 Binomial, Poisson, and Gaussian Distributions

In probability theory's simplest form, we try to identify mutually exclusive events for a problem, such as the outcome of tossing a die. Ideally, we expect that the likelihood of seeing any face of the six-sided die is equal and, in an experiment, we equate the probability of obtaining, say, a 4 with the frequency of such events relative to the total number of tosses. In the limit of a large number of trials, we expect that this ratio will assume a limiting value of 1/6—this expectation is referred to as the "law of large numbers." This point needs to be emphasized since there are many misconceptions and myths present in common parlance relating to probability. Some individuals cite the false "law of averages" as guaranteeing that their favored football team, which has shown a losing streak, will begin to win games: That notion is simply wrong, and their observation can most simply be explained by the fact that their favored team is composed of a poorly coordinated and ineffectual group of players! Let us begin with the problem associated with a random process that has N possible outcomes, x_i for $i = 1, \ldots, N$. For tossing a single die, $N = 6$. The probability of obtaining the ith outcome is p_i for $i = 1, \ldots, N$. For the case of a die, each of the $p_i = 1/6$. It follows, since we have embraced in our accounting all possible outcomes, that

$$\sum_{i=1}^{N} p_i = 1. \tag{5.1}$$

We define the *mean* or average of such a process with outcomes x_i by

$$\langle x \rangle = \sum_{i=1}^{N} x_i p_i. \tag{5.2}$$

For the case of throwing a die, $x_i = i$ for $i = 1, \ldots, 6$ and $\langle x \rangle = 3\frac{1}{2}$. Note that the average outcome in this case does not correspond to a physically realizable state! It is common to see the symbol μ employed to designate $\langle x \rangle$, and the symbol E is employed to designate the operator corresponding to averaging, the so-called *expectation value*. Hence,

$$E(x) = \langle x \rangle = \mu. \tag{5.3}$$

As a measure of the variability of the outcomes, we introduce the *variance* of the process via the operator Var defined by

$$\mathrm{Var}(x) = E[(x - \mu)^2] = E(x^2) - 2E(x)\mu + \mu^2 = E(x^2) - \mu^2, \tag{5.4}$$

since the expectation or expected value of a constant is that constant. It is common practice to employ the symbol σ^2 for the variance, namely,

$$\sigma^2 = \mathrm{Var}(x). \tag{5.5}$$

In the case of a single die, it is easy to show that the variance $\sigma^2 = 35/12 \approx 2.916667$.

Let us digress briefly and consider randomness in one of its simplest guises, particularly as it affects observations and measurements. Suppose we have a quantity that is subject to some form of measurement error, the nature of which does not change, so that its statistical properties remain *stationary*. Consider repeated measurements of a quantity x_i, where i refers to the measurement taken, say, for $i = 1, \ldots, n$ where

$$x_i = \mu + \epsilon_i, \tag{5.6}$$

where ϵ_i describes the error. Further, we assume that

$$E(\epsilon_i) = 0, \tag{5.7}$$

and that it is uncorrelated so that

$$E(\epsilon_i \epsilon_j) = \delta_{ij} \sigma^2. \tag{5.8}$$

Accordingly, given our simple model (5.6), we have

$$E(x_i) = \mu$$
$$E[(x_i - \mu)^2] = \mathrm{Var}(x_i) = \sigma^2. \tag{5.9}$$

Suppose, then, that we seek to estimate the values of μ and σ^2. It follows immediately that the estimate $\bar{x}$ given by

$$\bar{x} = \frac{1}{n} \sum_{i=1}^{n} x_i = \mu + \frac{1}{n} \sum_{i=1}^{n} \epsilon_i \tag{5.10}$$

is *unbiased* since

$$E(\bar{x}) = \mu. \tag{5.11}$$

Surprisingly, we can show that

$$s^2 = \frac{1}{n-1} \sum_{i=1}^{n} (x_i - \bar{x})^2 \tag{5.12}$$

is an unbiased estimator for the variance. This expression may appear to be surprising inasmuch as we divide the summation by $n - 1$ rather than n. The reason for this is that $\bar{x}$ is not without error since it is based on existing measurements of x_j and their corresponding errors ϵ_j. The proof of this outcome is left as an exercise for the reader. We have provided these definitions (5.11) and (5.12) as a guide for estimating the mean and variance of observational data. When we turn to issues surrounding regression, we will observe that measures of uncertainty there also require that the number of observations n be reduced by the number of assumed parameters m. The problem that we have just reviewed, which has a single parameter (μ), is a case in point.

Let us expand the ideas above by exploring a version of the coin-toss problem generally known as the *drunkard's walk* (Chandrasekhar, 1943) that lies at the heart of Brownian motion and so-called Brownian walks. Imagine the plight faced by a drunkard on a street with lighted lampposts every ℓ in distance, and he wants to go home. The drunkard's sense of direction has vanished, and he walks, with equal likelihood, to the lamppost on his right or his left, whereupon he briefly collapses. We characterize each of his sojourns by the random variable ε_i, where i refers to the step he has just taken and $\varepsilon_i = \pm 1$. The time interval between each step is τ, and we follow his progress as he takes n steps. We observe that his position after the nth step is x_n given by

$$x_n = \sum_{i=1}^{n} \varepsilon_i \ell, \tag{5.13}$$

and observe, by definition, that

$$E(\varepsilon_i) = 0$$
$$E(\varepsilon_i \varepsilon_j) = \delta_{ij}. \tag{5.14}$$

Therefore, it follows that

$$E(x_n) = \sum_{i=1}^{n} E(\varepsilon_i)\ell = 0$$
$$E(x_n^2) = \sum_{i=1}^{n} \sum_{j=1}^{n} E(\varepsilon_i \varepsilon_j) = n\ell^2. \tag{5.15}$$

We identify the square root of this latter expression for $E(x_n^2)$ with the distance d_n traversed over n steps in time t_n, namely,

$$d_n = E(x_n^2) = \sqrt{n}\ell = \sqrt{\frac{t_n \ell^2}{\tau}}. \tag{5.16}$$

This result parallels earlier ones we obtained for diffusive processes using dimensional analysis and the heat equation, respectively, to identify the distance scale d affected over a time t, or $d \propto \sqrt{Dt}$ where D was the diffusion coefficient. Apart from a factor of 2, we can now identify D with ℓ^2/τ. An essential take-home lesson from this derivation is that deterministic processes generally scale in proportion to n but random ones scale as $\sqrt{n}$. The random quantity encountered here, namely, x_n, is considered *nonstationary* since its properties are dependent and change with time n.

Before we explore some of the most common distribution functions, let us discuss an issue that emerges when various outcomes are not statistically independent, and gives rise to *Bayes's theorem* (Feller, 1968). We shall address this using a simple example. Suppose that A corresponds to the proposition that an individual is a citizen of Spain and that B corresponds to the proposition that an individual speaks Spanish. We employ the symbol "$\cap$" to describe the intersection of these two sets, that is, that an individual is both a citizen of Spain and a speaker of Spanish. We can write, accordingly, that

$$P(A \cap B) = P(A|B)P(B), \qquad (5.17)$$

which may be read as saying that the probability that an individual is both a citizen of Spain and a speaker of English is the probability that an individual is a citizen of Spain *given* that the individual is a speaker of Spanish times the probability that an individual is a speaker of Spanish. Likewise, we can write

$$P(A \cap B) = P(B|A)P(A), \qquad (5.18)$$

which may be read as saying that the probability that an individual is both a citizen of Spain and a speaker of English is the probability that an individual is a speaker of Spanish *given* that the individual is a citizen of Spain times the probability that an individual is a citizen of Spain. For obvious reasons, probabilities of the form $P(A|B)$ are referred to as *conditional probabilities*. We can now manipulate these expressions to obtain explicitly these conditional probabilities, namely,

$$P(A|B) = \frac{P(A \cap B)}{P(B)}$$

$$P(B|A) = \frac{P(A \cap B)}{P(A)}. \qquad (5.19)$$

As an illustration, the first of these constructions may be read as saying that the probability that an individual is a citizen of Spain given that the individual is a speaker of Spanish is the probability that an individual is both a citizen of Spain and a speaker of Spanish divided by the probability that an individual is a citizen of Spain. The logic employed here is general and very much independent of the specific example, and is widely used to simplify the calculation of probabilities in many complicated problems. We turn now to derive some of the more widely employed distribution functions.

5.1.1 Binomial Distribution

Consider now an experiment where one specific outcome is designated a "success" and all others are designated as "failures." If the probability of success is p, then it follows that the probability of failure, q, is $1 - p$. Moreover, let us now consider N repetitions of the experiment where each successive outcome is independent of the previous result. For example, if we were drawing lottery tickets from a basket in sequence, but returned each drawn ticket to the basket, the probability of obtaining a given ticket is independent of the sequence in which it was drawn. A natural question to ask, then, is what is the probability of obtaining m successes and $N - m$ failures, where $m = 0, \ldots, N$? This is often called a *binomial* or *Bernoulli* trial (Feller, 1968), and that probability, which we will designate as $P(m; N, p)$, using the usual definition of a combination

$$_N C_m \equiv \binom{N}{m} = \frac{N!}{m!(N-m)!}, \tag{5.20}$$

is given by (Moran, 1968)

$$b(N, p, m) = \binom{N}{m} p^m q^{N-m}. \tag{5.21}$$

The latter expression emerges since there are $_N C_m$ ways of obtaining m successes and $N - m$ failures having independent probabilities each of p and of q, respectively. Finally, we note that

$$\sum_{m=0}^{N} b(N, p, m) = \sum_{m=0}^{N} \binom{N}{m} p^m q^{N-m} = (p + q)^N = 1, \tag{5.22}$$

since the summation includes the totality of all possible outcomes. We wish to calculate the mean and variance for this distribution and observe, first, that

$$\mu = \sum_{m=0}^{N} m \binom{N}{m} p^m q^{N-m} = Np, \tag{5.23}$$

and that

$$\sigma^2 = \sum_{m=0}^{N} m^2 \binom{N}{m} p^m q^{N-m} - \mu^2 = Npq = Np(1-p), \quad (5.24)$$

after some algebra.

As a simple illustration of these results, we will ask, what is the likelihood in a repeated tossing of a die of obtaining no 1's in 6 tosses? Since a success here corresponds to getting a result other than a 1, $p = 5/6$ and the probability of getting no 1's (or any other set outcome) is $(5/6)^6 \approx 0.401878$. Returning to the fallacious "law of averages," assuming that you observed no 1's in 6 trials, what is the likelihood that the seventh trial would yield a 1?—the result is $1/6$ since the outcome is independent of all previous die tosses.

5.1.2 Poisson Distribution

A commonly encountered situation is one where $p \ll 1$ and $N \gg 1$ but pN is moderate. This problem emerges in many situations where the probability of a success is low but there are many repetitions. We wish to explore the binomial probability result (5.20) in that limit but where $\mu = Np$ is neither very small nor very large. In other words, what is the probability of a rare event in circumstances where many opportunities prevail for it to happen? We observe, for that circumstance and where $m \ll N$, that

$$b(N, p, m) = \binom{N}{m} p^m q^{N-m}$$

$$\approx \frac{N \cdot (N-1) \cdot \cdots \cdot (N-m+1)}{m!} p^m (1-p)^{N-m}$$

$$\approx \frac{N^m}{m!} p^m (1-p)^N \approx \frac{\mu^m}{m!} (1-p)^{Np/p}$$

$$\approx \frac{\mu^m}{m!} \exp(-Np) = \frac{\mu^m}{m!} \exp(-\mu), \quad (5.25)$$

where we have made use of the identity

$$\lim_{p \to 0} (1-p)^{1/p} = \exp(-1). \quad (5.26)$$

We will refer to this limiting case as the *Poisson distribution* $P(\mu, m)$, which describes the probability of having m successes in a random process where the mean value is μ. Further, we observe that

$$\sum_{m=0}^{\infty} P(\mu, m) = \sum_{m=0}^{\infty} \frac{\mu^m}{m!} \exp(-\mu) = 1, \qquad (5.27)$$

noting that this incorporates the usual power series expansion for $\exp(\mu)$. In like fashion, we can show that

$$E(m) = \sum_{m=0}^{\infty} mP(\mu, m) = \mu, \qquad (5.28)$$

and that

$$E(m^2) = \sum_{m=0}^{\infty} m^2 P(\mu, m) = \mu(\mu + 1) \qquad (5.29)$$

so that

$$\text{Var}(m) = \mu. \qquad (5.30)$$

One final result we wish to show here pertains to the likelihood of not having any successes, that is, $m = 0$, in circumstances where the number of trials N is $1/p$, resulting in $\mu = 1$. It immediately follows that the probability of no successes is $\exp(-1) \approx 0.367878$ which compares with our earlier calculation for never getting a 1 in six throws of a die. Hence, the probability of at least one success is $1 - \exp(-1) \approx 0.632121$. We have, therefore, invalidated a well-known urban myth: if the odds of winning something is p and you gamble $1/p$ times, you only win some 63% of the time.

A special but important case of a Poisson distribution or Poisson process emerges in continuous processes where the probability p refers to the probability of an event occurring in one unit of time. Therefore, over a time interval t, the mean number of events μ can be expected to be pt. As an illustration of this relationship, the probability over time t of no occurrences (i.e., $m = 0$) becomes $\exp(-pt)$. This formula is the same as that employed in our discussion relating to radioactive decay, where p corresponds to the decay rate y that we introduced in Eq. (4.40). For *in situ* radioactive decay, the number of parent nuclei $n(t)$ decays according to

$$\frac{dn(t)}{dt} = -yn(t), \qquad (5.31)$$

whose solution is

$$n(t) = \exp(-yt)n(0), \qquad (5.32)$$

thereby showing the connection with the Poisson distribution we have derived for discrete events and the distribution associated with continuous events. This demonstrates the connection between the Poisson distribution and the inter-arrival times of decay products.

We now want to extend these results to the realm where m is in the neighborhood of $\mu = Np$, taking us to the so-called *normal* or *Gaussian* distribution. In so doing, we will make a transition from discrete to continuous distribution functions.

5.1.3 Normal Distribution

Following Moran (1968), we introduce Stirling's approximation (3.30) into our expression (5.20)

$$b(N,p,m) = \binom{N}{m} p^m q^{N-m}$$

$$\approx \frac{1}{\sqrt{2\pi}} \frac{N^{N+1/2}}{m^{m+1/2}(N-m)^{N-m+1/2}} p^m q^{N-m}$$

$$\approx \sqrt{\frac{N}{2\pi m(N-m)}} \left(\frac{Np}{m}\right)^m \left(\frac{Nq}{N-m}\right)^{N-m}. \qquad (5.33)$$

Let us now define

$$x = m - Np, \qquad (5.34)$$

and, recalling for small x, that is, $|x| < Npq$, that

$$\ln(1 + x) \approx x - \frac{x^2}{2} + \cdots, \qquad (5.35)$$

we can write

$$\ln\left[\left(\frac{Np}{m}\right)^m \cdot \left(\frac{Nq}{N-m}\right)^{N-m}\right]$$

$$= -m \ln\left(\frac{m}{Np}\right) - (N-m)\ln\left(\frac{N-m}{Nq}\right)$$

$$= -m \ln\left(1 + \frac{x}{Np}\right) - (N-m)\ln\left(1 - \frac{x}{Nq}\right)$$

$$\approx -m\left[\frac{x}{Np} - \frac{x^2}{2(Np)^2}\right] + (N-m)\left[\frac{x}{Nq} - \frac{x^2}{2(Nq)^2}\right]$$

$$= -\frac{x^2}{N}\left(\frac{q+p}{qp}\right) + \frac{x^2}{2Npq} = -\frac{x^2}{2Npq} = -\frac{x^2}{2\sigma^2}, \qquad (5.36)$$

which establishes that

$$\left(\frac{Np}{m}\right)^m \left(\frac{Nq}{N-m}\right)^{N-m} = \exp\left(-\frac{x^2}{2\sigma^2}\right). \qquad (5.37)$$

Finally, we observe that

$$\frac{m(N-m)}{N} = \frac{(x+Np)(N-x-Np)}{N}$$

$$\approx \frac{(Np+x)(Nq-x)}{N}$$

$$\approx Npq = \sigma^2. \qquad (5.38)$$

We can now pull all of these results together to observe that

$$b(N,p,m) \approx \frac{1}{\sqrt{2\pi\sigma^2}} \exp\left(-\frac{x^2}{2\sigma^2}\right), \qquad (5.39)$$

thereby verifying in the appropriate limit of $|m - Np| \ll Npq$ that the binomial distribution becomes the normal or Gaussian distribution. As a consequence of the genericness of the binomial distribution, we expect that the normal distribution will commonly appear in real-world applications. It is quite common to find that experimental error is normally distributed.

We now regard x as a continuous variable, and we become concerned with the integral over a range of x in assessing whether an observation is consistent with a normal distribution. In Figure 5.1, we present the normal distribution typically identified as $\mathcal{N}(x; \mu, \sigma)$ for $\mu = 0$ and $\sigma = 1$.

We identify by horizontal lines terminating in arrowheads the regions corresponding to the probability of an event being within 1, 2, and 3 standard deviations. It is noteworthy that these three measures of departure from the mean correspond to the standards employed in the social sciences (1 standard error or 68%), the life sciences (2 standard errors or 95%), and the physical sciences (3 standard errors or 99.7%). The significance of these measures are often misunderstood. A simple thought experiment highlights how these ideas are commonly misused.

Consider conducting a poll with N respondents, with two choices available, and you wish to assess the significance of the outcome, that is, how likely one of the choices will prevail over the other in, for example, an election. Let us employ

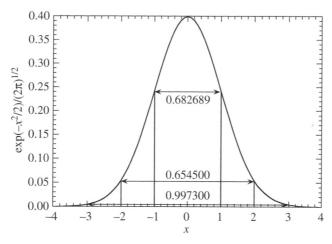

Figure 5.1. Gaussian profile displaying probabilities of events within 1, 2, and 3 standard errors.

as an analogy a coin toss experiment where a head is given a score of $+1$ and a tail is given a -1. From our discussion earlier of a drunkard's walk, the uncertainty in the outcome is *on average* $\sqrt{N}$, making the relative uncertainty $1/\sqrt{N}$, which is often expressed as a percentage. It is vital to remember that this is the uncertainty that should be expected. For example, if $N = 1600$, a typical number of samples taken in telephone polls by major news organizations, the uncertainty is generally cited as $100/\sqrt{1600} = 2.5\%$. This corresponds to one standard error, and news organizations often make the mistake of claiming that differences larger than that are essentially impossible. However, we observe from our plot of the normal distribution that departures of 32% are to be expected (16% in each direction). What is remarkable is how this simple misunderstanding of statistics, plus other influences, has led to the wholesale failure of predictions in recent American elections (Silver, 2012). Perhaps the best known of these was the 2000 presidential election where polling performed by various organizations in the days leading up to the election suggested that Al Gore was ahead of George Bush by about one standard error and, based on that observation, erroneously predicted a Gore victory—the political pundits did not appreciate that a difference of one standard error made the race wide open. Freedman et al. (2007) elaborate on this issue, showing how the misapplication of statistics can have profound consequences. We now turn our attention to understanding why in a broader arena of problems the normal distribution is prevalent.

In all matters of science, we can never prove that a conjecture or hypothesis is true, only that it is false. This reflects itself in the application and interpretation of statistics. Thus, while we cannot prove the validity of a conjecture, we can show that a seemingly opposite conjecture—known as a *null hypothesis*—fails at some level of confidence. Consider a simple example: We toss a coin 10 times and obtain 10 heads. The hypothesis here, designated H_0, is that the coin or the manner in which it was tossed was not "fair." The null hypothesis, designated H_1, would be that the coin and its tossing were "fair." Using the binomial distribution, we can immediately show that the probability of getting 10 consecutive heads is $2^{-10} \approx 0.1\%$. Therefore, we reject the null hypothesis at the 0.1% level. Closely related to this methodology is the use of *p values*; in our example, the *p* value was 0.1%. Typical *significance levels* for rejection are the 5% and 1% values; our example here is much more compelling. Since our *p* value is much smaller than the significance level, we reject the null hypothesis. This, though, is not the same as accepting our original hypothesis—only the null hypothesis, if appropriately constructed, can be rejected.

5.2   *Central Limit Theorem*

In the preceding section, we have made the transition from discrete events to a continuum approximation. In essence, the numbers of samples associated with our derivation of the normal distribution is so large that summations can be replaced by integrals with discrete intervals being identified relative to some very large number, rendering those intervals infinitesimal. In many instances, we can identify for a continuous random variable x the likelihood of an event $p(x)\,\mathrm{d}x$ between x and $x + \mathrm{d}x$. We refer to $p(x)$ as a *probability distribution function*. For a variety of technical but sometimes significant reasons, mathematicians prefer to employ its integral $P(x)$, known as a *cumulative distribution function*, defined according to

$$P(x) = \int_{-\infty}^{x} p(x')\,\mathrm{d}x', \qquad (5.40)$$

where we have assumed that $-\infty < x < \infty$. (Finite limits can be employed with only minor modifications in the argument provided.) Since we are dealing with probabilities, which in turn relate to frequencies of events, we can assert

$$\frac{dP(x)}{dx} = p(x) \geqslant 0$$

$$\int_{-\infty}^{\infty} p(x)\,dx = 1$$

$$\int_{-\infty}^{\infty} xp(x)\,dx = \mu$$

$$\int_{-\infty}^{\infty} (x - \mu)^2 p(x)\,dx = \sigma^2. \tag{5.41}$$

As a general rule when these conditions are met, the Fourier transform or so-called *characteristic function* exists and we define it and its inverse according to (Feller, 1968)

$$\tilde{p}(k) = \int_{-\infty}^{\infty} p(x)\exp(ikx)\,dx = E[\exp(ikx)]$$

$$p(x) = \frac{1}{2\pi} \int_{-\infty}^{\infty} \tilde{p}(k)\exp(-ikx)\,dk. \tag{5.42}$$

We have derived in chapter 3 that the Fourier transform of a Gaussian is itself a Gaussian.

Accordingly, we note that

$$\hat{p}(k) = \exp\left(-\frac{k^2\sigma^2}{2} + ik\mu\right) = \int_{-\infty}^{\infty} p(x)\exp(ikx)\,dx$$

$$= \int_{-\infty}^{\infty} \frac{1}{\sqrt{2\pi\sigma^2}} \exp\left[-\frac{(x-\mu)^2}{2\sigma^2} + ikx\right]dx. \tag{5.43}$$

Also for the Gaussian case, we observe that

$$P(x) = \mathcal{N}(x; \mu, \sigma) = \frac{1}{2}\,\mathrm{erfc}\left[\frac{-x+\mu}{\sqrt{2\sigma^2}}\right], \tag{5.44}$$

where $\mathrm{erfc}(x)$ designates the complementary error function $1 - \mathrm{erf}(x)$ in x (Olver, 2010). We will now employ these results to derive the *central limit theorem*.

For convenience, let us now define a function $f(k)$ for a given $p(x)$ that is closely related to the characteristic function $\tilde{p}(k)$ according to

$$f(k) \equiv \int_{-\infty}^{\infty} p(x)\exp\left[ik\frac{x-\mu}{\sigma}\right]dx. \tag{5.45}$$

Consequently, we note, where a prime indicates differentiation with respect to k, that

$$f(0) = \int_{-\infty}^{\infty} p(x)\,dx = 1$$

$$f'(0) = \int_{-\infty}^{\infty} p(x)i\frac{x-\mu}{\sigma}\,dx = 0$$

$$f''(0) = -\int_{-\infty}^{\infty} p(x)\left[\frac{x-\mu}{\sigma}\right]^2\,dx = -1, \tag{5.46}$$

allowing us to approximate $f(k)$ by

$$f(k) \approx 1 - \frac{k^2}{2}. \tag{5.47}$$

Suppose, now, that we are dealing with a random variable X that is a particular linear combination of $x_1, x_2, \ldots, x_n$, namely,

$$X = \frac{1}{\sqrt{n}\sigma} \sum_{i=1}^{n} (x_i - \mu). \tag{5.48}$$

We can now exploit the characteristic function $\tilde{p}(k)$ to obtain the probability distribution function corresponding to X, which we shall call $p(X)$. We observe that

$$\tilde{p}(k) = \int_{-\infty}^{\infty} p(X)\exp(ikX)\,dX = E[\exp(ikX)]. \tag{5.49}$$

We utilize our definitions for X and $f(k)$ to obtain

$$\tilde{p}(k) = \int_{-\infty}^{\infty} \cdots \int_{-\infty}^{\infty} \exp\left[\frac{1}{\sqrt{n}\sigma} \sum_{i=1}^{n}(x_i - \mu)\right]$$

$$\times p(x_1) \cdots p(x_n)\,dx_1 \cdots dx_n$$

$$= f\left(\frac{k}{\sqrt{n}}\right) \cdots \left(\frac{k}{\sqrt{n}}\right) \approx \left(1 - \frac{k^2}{2n}\right)^n \approx \exp\left(-\frac{k^2}{2}\right). \tag{5.50}$$

Finally, we invert the Fourier transform to obtain

$$p(X) = \frac{1}{2\pi} \int_{-\infty}^{\infty} \tilde{p}(k)\exp(-ikX)\,dk$$

$$\approx \frac{1}{2\pi} \int_{-\infty}^{\infty} \exp\left(-\frac{k^2}{2}\right)\exp(-ikX)\,dk$$

$$\approx \frac{1}{\sqrt{2\pi}} \exp\left(-\frac{X^2}{2}\right). \tag{5.51}$$

Thus, we have now proved for a very broad class of distribution functions that sums of those associated random variables are to a very good approximation normally distributed—this is the essence of the central limit theorem.

When summing different random variables suitably normalized, a rather similar result, often called the *Lindeberg–Feller theorem*, applies. Moreover, in many geophysical applications the presence of a hierarchical or fractal structure can give rise to variables that are combined not in the form of a sum but in the form of a product, generally, of positive-valued quantities. Using similar reasoning for the logarithm of the composite variable, which can now be expressed as the sum of the logarithms of its component parts, we can derive what is called a *log-normal distribution*. What is observed in a variety of problems is that the empirically derived distribution function that fits the observed data has a "long tail"; a log-normal distribution introduces an exponential of the square of a logarithmic variable and, hence, can persist longer than a power-law distribution—sometimes called a *Pareto* distribution. The problem area of long- or heavy-tailed distributions remains an active area of research.

5.3 Randomness in Data and in Simulations

Having introduced some essential features of probability theory and statistics, we seek here to address two classes of issues where randomness plays a fundamental role. In the first, random error often appears in experimental measurements originating both in the laboratory and in simulations. In the second, we introduce into simulations a random component describing the initial distribution of some aspect of the physical problem.

5.3.1 Regression and Fitting of Experimental Data

With the insights we have gained, we can now approach the issue of fitting models to observed data. We have already discussed one, the case where the observations result in a quantity together with additive noise such that the calculation of its arithmetic mean provided a good estimate of its value. Let us now consider the case of data fitting a straight line, say

$$y = mx + b. \tag{5.52}$$

If (x_i, y_i) for $i = 1, \dots, N$ are observations, then the *least-squares method*, nominally attributed to Gauss, can be expressed as finding the slope m and intercept b such that the sum

$$U = \sum_{i=1}^{N} (mx_i + b - y_i)^2 \qquad (5.53)$$

is a minimum. We accomplish this by taking the derivative of U with respect to both variables, thereby making U an extremum, namely,

$$\frac{\partial U}{\partial m} = 2 \sum_{i=1}^{N} x_i(mx_i + b - y_i) = 0$$

$$\frac{\partial U}{\partial b} = 2 \sum_{i=1}^{N} (mx_i + b - y_i) = 0. \qquad (5.54)$$

This yields the linear pair of equations, sometimes called the *normal form*,

$$m \sum_{i=1}^{N} x_i^2 + b \sum_{i=1}^{N} x_i = \sum_{i=1}^{N} x_i y_i$$

$$m \sum_{i=1}^{N} x_i + bN = \sum_{i=1}^{N} y_i. \qquad (5.55)$$

This pair of equations is then solved to yield an unbiased estimate of the unknowns. The measure of the noise that is present emerges from calculating the least squares total error U and dividing it by $N - 2$, the total number of degrees of freedom in the problem. This corresponds to the number of data points less the two parameters.

To illustrate this, we have generated some noise ϵ_i to introduce into our model, namely,

$$y_i = mx_i + b + \epsilon_i, \qquad (5.56)$$

where $m = 0.5$ and $b = 1$. The simulation results are shown in Figure 5.2.

While there are two points well above the line, they are counterbalanced by the 9 points below it. The goodness of fit is sometimes expressed using a *correlation coefficient*, which is calculated using the coefficients of the 2×2 matrix inverted in obtaining m and b. There are a broad range of methodologies employed in fitting data to theoretical curves, much of it originating in life-science applications. D'Agostino and Stephens (1986) is an excellent resource for addressing these issues.

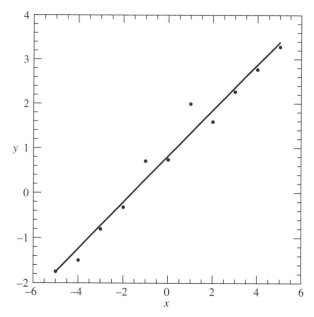

Figure 5.2. Least-squares regression fit to $y = 0.5x + 1$; coefficient estimates $m = 0.514$ and $b = 0.816$.

5.3.2 Random Number Generation and Monte Carlo Simulation

The final topics that we will address in this chapter are some of the probabilistic issues that emerge in developing pseudorandom number generators and the performance of Monte Carlo simulations. Inasmuch as randomness is ubiquitous in nature, the ability to create numerical experiments containing random contributions is essential, but there are a number of criteria that must be met. The methodology should be portable, that is, will rapidly yield the same results on different computers using different computational platforms, by different users. It should be deterministic, so that the user can repeat specific experiments, and should yield a sequence of numbers that are uncorrelated but approximate the properties of the distribution function of interest. A pseudorandom number generator is an algorithm for generating such a sequence. We refer to this generator as being "pseudorandom" because, while we maintain absolute control over what it does and it is fundamentally deterministic, the outcome of such a scheme should convince any observer that it appears to be random yet is drawn from a population consistent with a specified distribution function.

In our discussion of mappings, we identified Von Neumann's use of the circle map (2.199). This, historically, appears to have been among the first efforts at artificially generating random numbers. Knuth (1997b) provides a review of many such methods together with a detailed description of many such algorithms. Press et al. (1988) provides many "numerical recipes" for performing such manipulations including codes in Fortran and C; the benefit of such programs is that they have been thoroughly tested and shown to satisfy the essential criteria yet remain portable and highly reliable. In recent years, number theoretic approaches have yielded somewhat more rigorous methodologies, but these tend to be hard to implement and generally are much slower. There were others, of course, built on manipulations with integers—for example, the mid-square method where you take an integer with many digits, square it and extract the middle set of digits as a new number, and repeat—however, these tend to be very limited in utility. Almost all good schemes are integer based and exploit the ability of modern computers to perform many-digit integer arithmetic exactly, thereby allowing the formulation of pseudorandom sequences that do not repeat themselves for 10^8 executions or more. Among the best of these are so-called *linear congruential generators*: In a loose sense, the circle map can be regarded as an example of such a scheme. These are integer-based algorithms that employ a linear relation—hence, they are fast—and are readily implemented. Essentially all employ modulo arithmetic that is easy to implement in computer hardware. The sequence X_n is defined by a recurrence relation of the form

$$X_{n+1} = (aX_n + c) \bmod m, \qquad (5.57)$$

where m is a positive integer called the "modulus," the integer $a < m$ is the "multiplier," the integer $c < m$ is the increment, and the integer X_0 is the "seed" or "start value" (Press et al., 1988). If $c = 0$, this is sometimes called a multiplicative congruential generator. Press et al. (1988) provide a number of such generators, including verified computer codes, for eight-digit pseudorandom numbers that, when normalized, provide uniformly distributed random deviates over the interval $(0, 1)$ but do not show any periodicity over hundreds of millions of iterations.

Suppose now that we have implemented a scheme for producing uniformly distributed X_n. How then can we generate pseudorandom numbers corresponding to a cumulative distribution $P(Y)$? In essence, we make the association $X_n = P(Y_n)$, and then solve for Y_n as the random variable that we seek distributed according to $P(Y)$. For many distributions, this is relatively easy but is rather cumbersome with the normal distribution inasmuch as it involves inverting the error function. Box and Muller (1958) developed a simple scheme using logarithms, sines, and cosines for taking a pair of uniformly distributed random deviates, say X_n and X_{n+1}, and producing two normally distributed random deviates Y_n and Y_{n+1}.

Having surveyed some of the important issues emergent from considerations of probability and statistics in application to the geosciences, we conclude this chapter by discussing how the computer can be employed to explore geophysical problems.

5.4 Computational Geophysics

Many geophysical problems are fundamentally nonlinear and it becomes necessary to employ numerical methods to explore them. We begin by addressing issues limiting computational techniques resulting from approximate methods and computer round-off, as well as seminumerical algorithms that emerge from data mining considerations. We proceed to the solution of nonlinear equations, that is, the identification of roots of a single or set of nonlinear equations, a situation that commonly appears in geophysics. Finally, we turn to the numerical solution of ordinary and partial differential equations.

5.4.1 Computation, Round-off Error, and Seminumerical Algorithms

All computers provide a limited amount of accuracy, due to the number of bits present in their hardware. Most of today's computers have "double-precision hardware" affording 64 bits (employing binary values of 0 or 1) in their arithmetic processors. The protocol underwent a dramatic change with the introduction in the 1980s of the *IEEE 754* standard for performing floating-point arithmetic. Contemporary computers employ a form of scientific notation that has a sign bit, an 11-bit *exponent* in base 2, and a 53-bit *mantissa*. The first bit in this "fraction" or mantissa is understood to be 1, with 52 bits explicitly provided. The exponent of 2 has 1023 subtracted from it, allowing for a representation of numbers in decimal arithmetic from 10^{-308} to 10^{308}. Cody (1981) provides a historical treatment of the evolution of the IEEE 754 standard. Kincaid and Cheney (2009) and Higham (2002) provide detailed explanatory descriptions of this standard and computer arithmetic. The nature of floating-point operations, unlike strictly integer ones, is that error due to rounding or *round-off error* is essentially unavoidable. For example, in multiplying the floating-point representation for 1/3 by 3, we obtain a result equal to 1 minus an error associated with the least-significant bit position. With 52 bits available to represent a number—generally as many as 3 fewer due to the use of base-16 arithmetic—we have 16 decimal place equivalent accuracy. If round-off error accumulates in proportion to the number of operations n, then our computed results fundamentally lose all significance after 10^{16} operations if the accumulation of round-off error is systematic. If we organize the manner in which arithmetic is performed (Higham, 2002), then we can reduce the accumulation of round-off error in a manner reminiscent of a drunkard's walk, making it proportional to $\sqrt{n}$, instead. In due course, we will encounter the errors associated with computation—the errors emergent from taking finite numbers of terms, for example, in a Taylor series—which is referred to as *truncation error*. In many geophysical problems, there is a tradeoff between round-off and truncation error: Performing more computations to reduce the role of the approximations used in various expansions also increases the role of round-off error. These problems provide very real limits to the accuracy of results obtained as well as to the time taken to obtain them. It is not unusual for geophysical applications to push the capacity and capabilities of modern computation to their limit. We will have more to say on this problem in due course.

There are many problems that emerge in applications that have a seminumerical basis. For example, in many situations, we need to sort a long list of numbers (or other objects whose ordering can be given a numerical basis). If our list contains N entries, as in a telephone directory, then we need to make the order of N^2 comparisons. With, say, a million names, that implies a computational effort involving the order of 10^{12} comparisons.

Until recently, typical algorithms such as the "bubble sort" would recursively compare pairs of elements until what would be the last entry was identified and placed there, repeating the process with successively shorter lists. While reducing the computational effort by a factor of 2, the factor of N^2 remained. However, just as in the case of the FFT, we can seek to eliminate redundant comparisons. A simple device for achieving this emerges as follows: Regard your initial list as being composed of N lists of length 1. Compare the first list with the second, and produce a list with 2 elements; do the same for all such 1-member lists in your ensemble, and produce $N/2$ lists of length 2. Compare the first list (with two members) with the second, and produce a list with 4 elements; do the same for all such 2-member lists, and so on. Observe that in comparing lists, each with $m/2$ elements, only m operations are required in combining them and producing N/m lists with m elements. The end result is that the overall sorting process can be accomplished in $N \log_2 N$ operations. Knuth (1997a) provides a detailed analysis of a large class of methods that achieve this kind of efficiency; Press et al. (1988) provide codes for some of the most common implementations of this scheme, including Quicksort and Heapsort. These schemes can provide an additional speed enhancement of a factor of 2; however, the simple methodology we have just outlined will accomplish for large data sets monumental cost savings. With experience in programming, you will come to appreciate the need and value of focusing on the details needed to make your computations reliable, accurate, and cost effective. We turn now to one category of computation that has come to represent one of the nominally simplest yet most challenging problems in geophysics, the computation of the roots of equations.

5.4.2 Roots of Equations

Finding numerically the roots of equations is a common task in geophysics. This can occur in the context of evaluating special or transcendental functions, as in obtaining the roots of Bessel functions, or in solving high-degree polynomial equations that emerge in the dispersion relations for complex flows. Ralston and Rabinowitz (2001), Fröberg and Fröberg (1985), and Kincaid and Cheney (2009) provide surveys of many aspects of numerical analysis, including the numerical evaluation of roots to nonlinear equations. Polynomial equations are special in many ways, but also pose some of the most vexing problems. We have already reviewed the solution of polynomials through the third degree; fourth-degree polynomials can be reduced to cubics (Olver, 2010) and can also be solved in closed form. Fifth-degree polynomials with real coefficients must have at least one real root; we will describe shortly two absolutely convergent methods for finding one root. Then, that root can be factored—a process referred to as *deflation*—reducing the problem to fourth degree. The sources just mentioned, as well as Press et al. (1988), provide a brief introduction to a variety of special-purpose methods for finding all of the roots—often simultaneously—of high-degree polynomials. However, that task can be especially daunting due to round-off error and what becomes the intrinsically *ill-conditioned* nature of this problem. This issue also emerges in the algebraic eigenvalue problem associated with finding the eigenvalues of even moderate-sized matrices. For example, the notorious Wilkinson (1959) polynomial equation

$$f(x) = (x - 1)(x - 2) \cdots (x - 20), \qquad (5.58)$$

when expanded in the form $x^{20} - 210x^{19} + \cdots$ with the second coefficient changed to -210.0000001192 becomes a polynomial with the real parts dramatically altered, even in the first decimal place, with 10 of its 20 roots becoming complex conjugates.

Let us consider a pair of absolutely convergent methodologies for finding a root of any real $f(x)$ where you have identified two locations, say x_1 and x_2, such that

$$f(x_1)f(x_2) < 0; \qquad (5.59)$$

in words, these two estimates flank the exact solution. In the first method, that of *bisection*, we calculate the midpoint x_m according to

$$x_m = \frac{x_1 + x_2}{2}. \qquad (5.60)$$

In the unlikely event that $f(x_m) = 0$, we are done. Almost always, we then check to see if $f(x_m)f(x_2) < 0$. If so, we then replace x_1 by the value of x_m and repeat. Otherwise, we replace x_2 by the value of x_m and repeat. As is evident, the range of uncertainty (x_1, x_2) is reduced by a factor of 2 each time we go through this process—for every 10 such iterations, the range is reduced by three orders of magnitude. Methods where the uncertainty in an iteration is a fraction of the uncertainty in the previous one are said to exhibit linear *geometric convergence*. Other methods, which we shall discuss, where the uncertainty in one iteration is proportional to the square of the uncertainty in the previous one are said to exhibit *quadratic convergence*.

Another approach emerges from drawing a (dashed) secant line through $f(x)$ between those two points x_1 and x_2 and finding its intercept x_s according to

$$x_s = \frac{x_1 f(x_2) - x_2 f(x_1)}{f(x_2) - f(x_1)}. \tag{5.61}$$

This is generally referred to as the *secant method*; our implementation, with initial estimates flanking the root is called the *method of false position* or *regula falsi method* (Ralston and Rabinowitz, 2001). Figure 5.3 illustrates this process for $f(x) = 1 - \exp(-x)$ with our initial estimates being -1 and $+2$.

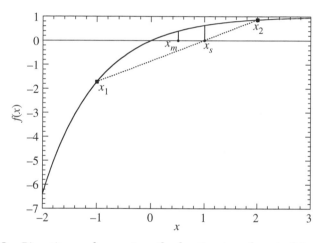

Figure 5.3. Bisection and secant method estimates of root of $1 - \exp(-x)$.

In situations where the second derivative of $f(x)$ is substantial, that is, where it shows significant curvature, the bisection method often works better. However, the secant method will ultimately provide more rapidly convergent results and, unlike the bisection method, which shows linear convergence, will become nearly quadratically convergent. Newton's method, which corresponds to replacing the secant line with a tangent line, is formally quadratically convergent, but in practice it is often divergent. For the sake of completeness, we show the estimate given by Newton's method x_N as

$$x_N = x_1 - \frac{f(x_1)}{f'(x_1)}. \tag{5.62}$$

It is evident, whenever the derivative of $f(x)$ is small, that estimates x_N of the solution can be very poor—indeed, unless substantial care is taken in identifying the initial root estimate, Newton's method often fails to converge. As a practical matter, alternating between the bisection method and the secant method (where the initial estimates flank the root) provides a highly efficient and absolutely convergent method of solution.

Finding simultaneous roots of more than one equation can sometimes be accomplished in the manner just described, but multidimensional versions of Newton's method often fail to converge. Suppose, for example, that we wish to find a solution (x, y) for the system

$$f(x, y) = 0$$
$$g(x, y) = 0. \tag{5.63}$$

As a rule, multidimensional root estimation is more difficult than one-dimensional root estimation. Imagine for a moment that x remains fixed in $f(x)$ and employ the methodology above to find a solution for y. In other words, you have established $y(x)$ in place of $f(x, y)$. You are in principle able to find a root of $g[x, y(x)]$ using this approach.

We will proceed momentarily to explore the application of numerical methods in the solution of ordinary and partial differential equations. There are many other categories of numerical problems that present themselves outside of differential equations, for example, linear algebra, eigenvalue analysis, and curve-fitting. Each of those topics are addressed in the references cited by Ralston and Rabinowitz (2001), Fröberg and Fröberg (1985), and Kincaid and Cheney (2009), and we will not explore these issues further here.

5.4.3 Numerical Solution of Ordinary Differential Equations

We will now develop the basis for finite difference methods using the prototypical equation

$$\frac{dx(t)}{dt} = f(x, t), \tag{5.64}$$

with initial condition $x(t_0) = x_0$; without loss of generality, we will presume that $t_0 = 0$. The description that we will provide here is valid not only for scalar equations, but also for a set of equations where we regard both x and f as "vectors." We introduce a time step Δt and estimated solution x_n such that, for $n = 0, \ldots,$

$$t_n = n\Delta t$$
$$x_n \approx x(t_n). \tag{5.65}$$

We are concerned with a triad of properties, sometimes referred to as consistency, accuracy, and stability (Richtmyer and Morton, 1994). The first of these is the requirement, in the limit $\Delta t \to 0$, that the finite difference expression reduces to the underlying differential equation. The accuracy criterion corresponds to the identification of the error term, the truncation error, present as a power in Δt. Finally, the stability criterion is the requirement for a solution that should be "nonincreasing," that is, decreasing or preserving its amplitude, that it does not grow in time. For analysis purposes, we will consider the linear paradigm

$$f(x, t) = -\gamma x, \tag{5.66}$$

where the real part of $\gamma > 0$ and has the solution

$$x(t) = \exp[-\gamma t]x(0). \tag{5.67}$$

We will presume, for simplicity, that γ is real during our remaining discussion. Gear (1971) as well as Hairer et al. (2009) explore these issues. LeVeque (2007) provides a good overview for the numerical solution of both ordinary and partial differential equations, while Ascher and Petzold (1998) is well suited for the nonspecialist. Note also that if f depends solely on t, without an explicit dependence upon x, then the solution to (5.64) corresponds to an integral—accordingly, we will briefly consider a variety of schemes for evaluating integrals.

We begin by examining the simplest integration scheme, due originally to Euler, that propagates explicitly the approximate solution forward in time with an error term that is formally $\mathcal{O}(\Delta t)^2$, namely,

$$x_{n+1} = x_n + \Delta t f(x_n, t_n) + \mathcal{O}(\Delta t)^2 = x_n + \Delta t f_n. \tag{5.68}$$

For the linear paradigm (5.66), it follows that

$$x_{n+1} = (1 - \gamma \Delta t)x_n + \mathcal{O}[(\Delta t)^2], \tag{5.69}$$

yielding the solution

$$x_{n+1} = (1 - y\Delta t)^n x_0. \tag{5.70}$$

We identify the latter expression as presenting a first-order approximation to the quantity $\exp(-ny\Delta t)$, thereby conforming with the need for consistency in the integration scheme. Moreover, we identify this as being second-order accurate in a local sense. Since we are concerned ultimately with the solution evaluated at some time T, for example, we identify there being $N = T/\Delta t$ steps taken in estimating $x(T)$ via x_N, thereby making the method order Δt accurate globally, that is, $\mathcal{O}[N(\Delta t)^2] = \mathcal{O}[\Delta t]$.

Importantly, we note that our formula (5.69) undergoes a radical change if $y\Delta t > 1$ and, especially, if $y\Delta t > 2$. In the first instance, we observe that x_n undergoes a sign change as n is incremented albeit, if $1 < y\Delta t \leqslant 2$, not increasing in amplitude with each time step. The first of these two circumstances conforms with our conceptual notion of stability, inasmuch as the estimated solution preserves its sign as it diminishes in amplitude. However, the second of these circumstances forms the basis mathematically for stability since the solution is nonincreasing. This displays a fundamental problem associated with explicit numerical integration schemes. We have focused on the scalar version of our paradigm (5.69), whereas the vector version would present a set of eigenvalues y. From a physical perspective, the largest eigenmode, corresponding to the behavior emergent from the largest eigenvalue, is often transient, that is, damps so rapidly as to be inconsequential for the duration of the calculation. However, its role in the underlying system of equations cannot readily be removed, and the stability constraint mentioned above, that is, $y\Delta t < 2$ for that eigenvalue, must be preserved. In many geophysical problems, the different rate constants present differ by orders of magnitude. Thus, in order to preserve the stability of the system of equations, a time step that is orders of magnitude smaller than the nominally dominant time scale may appear to be unavoidable. Gear (1971) provides a beautiful illustration of this phenomenon with two coupled equations where the eigenvalues differ by a factor of 100. We will refer to such systems as being *stiff*.

Let us consider another first-order solution scheme, which is *implicit*, also associated with Euler that has the form

$$x_{n+1} = x_n + \Delta t f(x_{n+1}, t_{n+1}) + \mathcal{O}(\Delta t)^2 = x_n + \Delta t f_n. \quad (5.71)$$

The essence of this being an implicit method is clear here: You need to know the answer to find the answer. Returning to our paradigm, we observe that

$$(1 + y\Delta t)x_{n+1} = x_n + \mathcal{O}[(\Delta t)^2], \quad (5.72)$$

whose solution is immediate and yields

$$x_n = (1 + y\Delta t)^{-n}x_0. \quad (5.73)$$

While comparable in consistency and accuracy with the result given by the explicit scheme (5.69), its solution is *always* stable. We say, therefore, that the explicit scheme (5.69) is conditionally stable whereas the implicit one (5.72) is unconditionally stable.

While on the issue of stability, it is important to understand that there are problems for which a given integration method is always unconditionally unstable. Consider the system, which corresponds to a simple harmonic oscillator $\ddot{x} + x = 0$,

$$\frac{dx}{dt} = +y$$
$$\frac{dy}{dt} = -x. \quad (5.74)$$

Writing the explicit first-order Euler method for this problem in matrix form, we observe that

$$\begin{bmatrix} x_{n+1} \\ y_{n+1} \end{bmatrix} \begin{bmatrix} 1 & -\Delta t \\ \Delta t & 1 \end{bmatrix} \begin{bmatrix} x_n \\ y_n \end{bmatrix}. \quad (5.75)$$

The eigenvalues for this matrix are $1 \pm i\Delta t$, which are larger than 1 in magnitude. Hence, the evolution of this system of equations results in the vector with components x_n and y_n undergoing both a rotation as well as growth. This simple problem illustrates how simple problems that otherwise present Hamiltonian features can become unconditionally unstable. Hamiltonian problems, which occupy a major role in geophysics and other disciplines, require special treatment and particular care (Sanz-Serna and Calvo, 1994; Hairer et al., 2010).

Now we turn to the issue of constructing integration schemes with greater accuracy. It should be evident that our first-order schemes suffered from approximating derivative information from the two endpoints of interest. One means of addressing this is to create an implicit scheme that incorporates information from both end points, as in

$$x_{n+1} = x_n + \tfrac{1}{2}[f(x_n, t_n) + f(x_{n+1}, t_{n1})]\Delta t. \qquad (5.76)$$

Note, in general, that this can be regarded as a nonlinear equation for x_{n+1} that must be solved using methods such as those described in the previous section. The linear paradigm offers an intriguing outcome, namely,

$$x_{n+1} = \left[\frac{1 - \gamma \Delta t/2}{1 + \gamma \Delta t/2}\right] x_n. \qquad (5.77)$$

In complex analysis, a transformation of this form is called a *linear fractional transformation* (Churchill, 1960) and, so long as $\mathrm{Re}(\gamma)\Delta t > 0$, the quantity inside the square brackets will remain smaller than 1 in magnitude, thereby guaranteeing stability for any Δt. This second-order implicit scheme not only overcomes many of the problems posed by intrinsically stiff systems, but renders the error locally to be third order, that is, $\mathcal{O}[(\Delta t)^3]$, and globally second order, that is, $\mathcal{O}[(\Delta t)^2]$. However, for nonlinear problems, it becomes necessary to solve a nonlinear system of equations at each step, which can reduce the efficiency of the process while helping to stabilize the integration scheme.

Another approach to obtaining second-order accuracy explicitly emerges from trying to estimate the solution at the midpoint, that is, $t_{n+1/2}$ and $x_{n+1/2}$, and using that information to generate are more accurate estimate at t_{n+1}. We establish two steps in achieving this objective:

$$x_{n+1/2} = x_n + \tfrac{1}{2}\Delta t f(x_n, t_n)$$
$$x_{n+1} = x_n + \Delta t f(x_{n+1/2}, t_{n+1/2}). \qquad (5.78)$$

This is sometimes referred to as a second-order (locally) *Runge–Kutta method*. It is easy to show that this methodology is stable for $\gamma \Delta t \leqslant 2$ and has a local error term $\mathcal{O}[(\Delta t)^3]$ and a second-order global error term.

Indeed, these concepts and methodology can be exploited to produce higher-order methods such as the widely used fourth-order Runge–Kutta method or RK4:

$$h_1 = f(x_n, t_n)$$
$$h_2 = f(x_n + \tfrac{1}{2}\Delta t h_1, t_{n+1/2})$$
$$h_3 = f(x_n + \tfrac{1}{2}\Delta t h_2, t_{n+1/2})$$
$$h_4 = f(x_n + \Delta t h_3, t_{n+1})$$
$$x_{n+1} = x_n + \tfrac{1}{6}\Delta t(h_1 + 2h_2 + 2h_3 + h_4). \qquad (5.79)$$

This method has a fifth-order error term locally and fourth order globally, hence its name. This class of schemes is referred to as a *single-step method* since it only requires knowledge of x_n at a single time step to proceed. *Multistep methods* tend to be more efficient since they exploit knowledge of former time step information to implicitly compute higher derivative terms in a *de facto* Taylor series. Hairer et al. (2009) provides a very up-to-date treatment of these schemes. Hairer et al. (2010) as well as Sanz-Serna and Calvo (1994) explores the role of geometry, which we identified in our description of the pendulum problem, and how it can be exploited to stabilize the evolution of nonlinear problems. Finally, while still on the topic of Runge–Kutta integrators, there is a widely used variant of this scheme called the RKF45 (Press et al., 1988) that calculates the local error term, allowing the user to control in an adaptive fashion the time step to bound the accumulated error. While widely used, the explicit Runge–Kutta RKF45 was observed to manifest problems and the DOPRI5 algorithm (Hairer et al., 2009) offers significant improvement. Step-size control is a very important topic inasmuch as we want to control the error produced yet at the same time make the calculation as efficient as possible. Ideally, we would like to be able to extrapolate to the limit $\Delta t \rightarrow 0$ and there are many schemes available for accomplishing this objective by suitable algebraic means, such as *Richardson extrapolation*.

This survey barely skims the surface of the very rich topic of numerical integration of ordinary differential equations. One subset of this problem corresponds to the computation of integrals that emerge when $f(x, t) \rightarrow f(t)$ is a function of time alone. In that situation, the distinction between explicit and implicit methods ceases to exists and the scheme (5.76) becomes the familiar *trapezoid rule*. We show in Figure 5.4 an illustration of the use of the trapezoid rule.

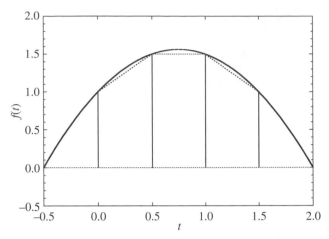

Figure 5.4.   Trapezoid rule for integrating $1 + \frac{3}{2}t - t^2$.

We wish to evaluate the integral of $f(t) = 1 + \frac{3}{2}t - t^2$ from $t = 0.0$ to $t = 1.5$; using (5.76), we see that it is approximated by the sum of the areas in the three trapezoids, each having a width of 0.5. Accordingly,

$$x(1.5) - x(0.0) \approx 0.25[f(0.0) + f(0.5)]$$
$$+ 0.25[f(0.5) + f(1.0)]$$
$$+ 0.25[f(1.0) + f(1.5)]. \qquad (5.80)$$

We also observe that the error in our expression corresponds to the mismatch between $f(t)$ and a set of interpolating chords spanning the range of integration and, therefore, is a consequence of the second derivative in $f(t)$ rendering the local error in our expression $\mathcal{O}[(\Delta t)^3]$ and the global error $\mathcal{O}[(\Delta t)^2]$. Moreover, because of symmetry, the higher-order error terms vary as even powers of Δt. Employing a procedure known as *extrapolation to the limit*, also known as *Romberg integration* (Press et al., 1988), we can systematically reduce the integration error in a highly efficient way. Suppose we consider the approximation obtained for the area as a function of Δt, calling it $A(\Delta t)$, and we refer to the exact result as $\mathcal{A}$. Since the global error is $\mathcal{O}[(\Delta t)^2]$, recalculating the integral via the formula (5.76) using a step size that is cut in half reduces the error by a factor of 4. (Attending carefully to detail, calculation of $f(t)$ at the same values of t can be avoided.) Therefore, we note that the expression

$$\mathcal{A} \approx \frac{4A(\Delta t/2) - A(\Delta t)}{4 - 1} = A(\Delta t/2) + \frac{A(\Delta t/2) - A(\Delta t)}{3} \qquad (5.81)$$

provides an improved result that is $\mathcal{O}[(\Delta t)^4]$. This turns out to be equivalent to *Simpson's rule* (Press et al., 1988) and fully incorporates contributions from $\dddot{f}(t)$.

We can systematize this process by creating a table whose first column is the values $A(\Delta t)$, $A(\Delta t/2)$, $A(\Delta t/4)$, and so on, where we stop when the difference between successive entries is sufficiently small for our purpose, that is, we have established convergence. Higher-order approximations can readily be generated by adding additional columns—see, for example, Fröberg and Fröberg (1985) or Press et al. (1988) for details—but there are two fundamental lessons emergent from this process. First, we must always check calculations performed to make certain that they have converged. Chaotic systems may preclude that, but we should be very certain that we are dealing with issues pertinent to sensitivity to initial conditions rather than ineffective implementation of a numerical algorithm. Second, we should exploit the emergent convergent properties, where possible, to improve the accuracy of the results that we obtain.

With these principles in mind, we proceed now to issues attendant to the numerical solution of partial differential equations.

5.4.4 General Issues in the Numerical Solution of Partial Differential Equations

We will focus once again on the application of the finite difference method in application to partial differential equations. There are many other kinds of methodologies available, including spectral methods (which exploit natural basis sets in the description of spatial variability) and finite element and related methods (which exploit the existence of underlying variational principles). We will not address issues emergent from elliptic equations, inasmuch as they fundamentally utilize the orthogonality properties of spherical harmonics and other orthogonal basis sets as we have described earlier. Details of implementation, nevertheless, are nontrivial and will be left to treatments dedicated to those classes of problems. Our focus here will be on parabolic and hyperbolic equations, which commonly emerge in geophysical applications. More detailed treatments of these are available in texts such as Potter (1973), Richtmyer and Morton (1994), and Peyret and Taylor (1990). LeVeque (2007) is a readily accessible all-round reference to the use of finite difference methods in the solution of ordinary and partial differential equations. We will employ a notation that is similar to that of the previous section, but that allows us to identify both time and space in a quantity, say, $u(x, t)$. In particular, we will write

$$u_j^n = u(x_j, t_n), \tag{5.82}$$

where $x_j = x_0 + j\Delta x$ and $t_n = t_0 + n\Delta t$. For convenience, it is useful to regard u at a given time t_n at all of its spatially sampled points x_j as a vector $\boldsymbol{u}$, thereby presenting us with a simple mode of expression of the problem, namely,

$$\frac{d\boldsymbol{u}}{dt} = \boldsymbol{Lu}, \tag{5.83}$$

where $\boldsymbol{L}$ describes an operator. This notation, as it happens, is at the heart of the *method of lines* (Schiesser, 1991), which facilitates the development of algorithms that are of high order in both space and time, but is beyond the scope of this treatment. Another ingredient that we wish to introduce into our analysis of numerical methods parallels Fourier approaches that we exploited when discussing partial differential equations. In particular, it is advantageous in many circumstances to be able to write an eigenmode $\exp(ikx)$ such that

$$u_j^n = u_0^n \exp(ikx_j). \tag{5.84}$$

We begin our exploration by investigating the diffusion equation.

5.4.5 Numerical Solution of Parabolic Partial Differential Equations

Let us now recall the diffusion equation in one dimension, namely,

$$\frac{\partial T(x, t)}{\partial t} = D\frac{\partial^2 T(x, t)}{\partial x^2}. \tag{3.88}$$

Replacing the time and space derivatives by finite differences, this expression (where u replaces T) can immediately be approximated by

$$u_j^{n+1} = u_j^n + \frac{D\Delta t}{(\Delta x)^2}(u_{j+1}^n - 2u_j^n + u_{j-1}^n) + \mathcal{O}[(\Delta t)^2, (\Delta x)^4]. \tag{5.85}$$

We note that, if we were to sum this expression over the index j, we would be in effect calculating the spatial integral over u, which we expect to be conserved from the underlying differential equation, and observe that the conservation law is maintained here. We illustrate in Figure 5.5 the structure of the mesh that we employ in solving the associated initial value problem using the equation above.

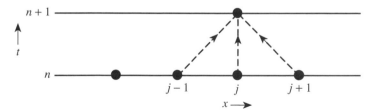

Figure 5.5. Diagram displaying space–time mesh employed in solving diffusion equation.

The first derivative in time was approximated by

$$\frac{\partial u}{\partial t} = \frac{u_j^{n+1} - u_j^u}{\Delta t}, \tag{5.86}$$

while the second derivative in time was approximated by

$$\frac{\partial^2 u}{\partial x^2} = \frac{u_{j+1}^n - 2u_j^n + u_{j-1}^n}{\Delta t}. \tag{5.87}$$

We now exploit the Fourier expression (5.84) to obtain

$$u_j^{n+1} = \left\{ 1 + \frac{D\Delta t}{(\Delta x)^2} [\exp(ik\Delta x) - 2 + \exp(-ik\Delta x)] \right\} u_j^n. \tag{5.88}$$

We refer to the term enclosed in braces as the *amplification factor*, namely,

$$g(\Delta t, \Delta x, D) = 1 + \frac{D\Delta t}{(\Delta x)^2} [\exp(ik\Delta x) - 2 + \exp(-ik\Delta x)], \tag{5.89}$$

since this term describes how the amplitude—as well as the phase—of a Fourier mode changes over one time step. After a little algebra, we observe for the finite difference scheme (5.87) that

$$g(\Delta t, \Delta x, D) = 1 - \frac{4D\Delta t}{(\Delta x)^2} \sin^2 \left(\frac{k\Delta x}{2} \right). \tag{5.90}$$

This expression reveals that both consistency and accuracy are assured if

$$\frac{4D\Delta t}{(\Delta x)^2} \ll 1, \tag{5.91}$$

and that stability is assured—that is, $|g| < 1$—if

$$\frac{4D\Delta t}{(\Delta x)^2} < 2, \tag{5.92}$$

since that will constrain the influence of the sinusoidal term. This stability result is generally written as

$$\Delta t \leqslant \frac{(\Delta x)^2}{2D}, \tag{5.93}$$

which has a simple explanation via dimensional analysis reasoning: In order for the method to remain stable, the time step Δt must be shorter than the time interval associated with diffusion extending over a distance greater than Δx.

There are many integration schemes available for the diffusion equation, and Richtmyer and Morton (1994) and Potter (1973) are excellent references. The methods we have now presented, however, can be employed to assess the consistency, accuracy, and stability of all of these. We now proceed to the advective equation as the simplest form of hyperbolic equation.

5.4.6 Numerical Solution of Hyperbolic Partial Differential Equations

Our motivation here is to develop an understanding of numerical approaches to solving hyperbolic equations in general, including the wave equation, but will focus on the simple advective equation

$$\frac{\partial u}{\partial t} + U \frac{\partial u}{\partial x} = 0, \tag{5.94}$$

where U is a velocity. The wave equation, in contrast, can be written

$$\frac{\partial^2 u}{\partial t^2} = c^2 \frac{\partial^2 u}{\partial x^2}; \tag{5.95}$$

We can employ a simple substitution to convert this wave equation into a pair of hyperbolic equations:

$$\frac{\partial u}{\partial t} = c \frac{\partial v}{\partial x}$$
$$\frac{\partial v}{\partial t} = c \frac{\partial u}{\partial x}. \tag{5.96}$$

In so doing, we have created a new variable $v(x, t)$ that can be established up to an additive constant since we know as an initial condition $\partial u(x, t)/\partial t$. The space–time mesh we introduced for parabolic equations remains useful, but we will shortly discover that stability considerations bring a few surprises!

Suppose that we discretize the advective equation (5.94) the same way we did the diffusion equation, namely,

$$u_j^{n+1} = u_j^n - \frac{U\Delta t}{2\Delta x}(u_{j+1}^n - u_{j-1}^n). \tag{5.97}$$

Employing the Fourier analysis procedure for obtaining the amplification factor that we introduced before, we obtain

$$g(\Delta t, \Delta x, U) = 1 - \mathrm{i}\frac{U\Delta t}{\Delta x}\sin(k\Delta x), \tag{5.98}$$

which we immediately identify with an amplitude and a phase. The latter corresponds to the phase/group velocity in the system, while the amplitude is seen to be larger than 1, rendering this explicit method unstable. One possible approach to stabilizing this methodology is to identify a means of reducing the 1 in the last expression so that the amplification factor would reside below 1. Lax (Potter, 1973) proposed to replace u_j^n by $\frac{1}{2}(u_{j+1}^n + u_{j-1}^n)$. Therefore, we employ the explicit scheme

$$u_j^{n+1} = \tfrac{1}{2}(u_{j+1}^n + u_{j-1}^n) - \frac{U\Delta t}{2\Delta x}(u_{j+1}^n - u_{j-1}^n), \tag{5.99}$$

which yields an amplification factor of

$$g(\Delta t, \Delta x, U) = \cos(k\Delta x) - \mathrm{i}\frac{U\Delta t}{\Delta x}\sin(k\Delta x). \tag{5.100}$$

We immediately observe for the modulus squared of the amplitude factor that

$$g(\Delta t, \Delta x, U) \cdot g^*(\Delta t, \Delta x, U) = 1 - \sin^2(k\Delta x)\left[1 - \left(\frac{U\Delta t}{\Delta x}\right)^2\right]. \tag{5.101}$$

We observe, so long as

$$\left(\frac{U\Delta t}{\Delta x}\right)^2 \leqslant 1, \tag{5.102}$$

that the solution is stable. Thus, we observe that

$$|U\Delta t| < \Delta x, \tag{5.103}$$

which is known as the *Courant–Friedrichs–Lewy condition*, is sufficient to assure stability. Equivalently, the solution will be stable as long as the time step Δt is sufficiently short to assure that a perturbation cannot propagate beyond the distance Δx, the spatial distance between mesh points.

The topic of the numerical solution of partial differential equations remains a cornerstone of modern geophysics. In developing methods of solution, we continue to focus on issues such as consistency, accuracy, and stability—as in the case of ordinary differential equations—but also seek to exploit any conservation laws that may be present in the underlying equations and physics. In addition to dealing with issues having to do with programming languages such as FORTRAN 95, C++, and Python as well as high-level packages such as Matlab, Mathematical, and Maple, we have entered an era of massive parallel computation with computers composed of tens of thousands of processors. New methodologies will necessarily evolve to fully exploit these capabilities, and our ability to synthesize methods of mathematical analysis, Monte Carlo and other statistical approaches, and advanced methods of computational investigation is quickly taking geophysics to a higher level of sophistication and capabilities.

5.5 Exercises

1. For a single die, show that the variance is $35/12$. For a pair of dice, calculate the mean and the variance.

2. What are the mean and the variance for the single die-toss problem?

3. Show that the estimator $\bar{x}$ for μ described in (5.11) is unbiased.

4. Show that the estimator s^2 for σ^2 described in (5.12) is unbiased. [Hint: Express $\bar{x}$ using a sum in x_j, and evaluate the expectation value of all terms that arise.]

5. If an election were to be held tomorrow and polling results with 900 respondents were available and 55% of those polled favored one candidate over the other, how certain would you be of the outcome?

6. Calculate the mean and variance for the Poisson distribution.

7. Derive the Fourier transform of the Gaussian transform (5.43).

8. Digital computers are finite state devices. They can represent a finite number of integers limited by 2^N, where N is the number of bits available in hardware. Accordingly, show why all linear congruential generators must ultimately be periodic. [Hint: Suppose each recurrence of the generator is distinct for 2^N operations, the maximum possible—what will happen with the next operation?]

9. The exponential distribution has a probability distribution $p(y) = \exp(-y)$. Obtain its cumulative distribution function $P(y)$. Derive a formula for obtaining an exponentially distributed random variable y from an associated uniformly distributed random variable x.

10. Using a calculator, use the method of false position to find inside $(0, 6)$ any root of

$$f(x) = (x - 5)(x - 2)(x - 1);$$

call that root estimate x_0. Deflate the polynomial $f(x)$ by dividing it by $x - x_0$, and show the second-order polynomial that emerges. Finally, solve the quadratic equation that emerges for the other roots. Note: All calculations should be performed numerically. This is a numerical exercise, not an analytic one.

11. Suppose that

$$f(x, y) = x + y - 1$$

and

$$g(x, y) = x^2 - \exp(y).$$

Find a real-valued solution to this pair of equations. Can other solutions to this problem exist?

12. Consider the coupled equations (5.74) for a simple harmonic oscillator, and define z by $x + iy$. Show that

$$\frac{dz}{dt} = -iz.$$

Accordingly, show that the "mid-point" integration formula (5.76) is stable.

13. Employ the second-order Runge–Kutta (5.78) and show that the solution is unconditionally unstable. Calculate the

effective growth rate of $|z|$. This question together with the previous one illustrates that simple Hamiltonian equations—Newton's laws—can pose special difficulties.

14. Consider the integrand shown in Figure 5.4 where we employed the trapezoid rule. Evaluate explicitly using the trapezoid rule using one interval from 0 to 1.0, and then do the same breaking the domain into two intervals going from 0 to 0.5 and from 0.5 to 1.0. Using the Romberg process once—Eq. (5.81)—evaluate the integral using what is called Simpson's method. Evaluate the integral analytically, since the integrand is very simple. Compare your analytic result with that due to Simpson's rule.

15. Suppose you integrate using the explicit parabolic finite difference method (5.85) the temperature distribution u between $x = 0$ and $x = 1$, where $\Delta x = 0.1$ and where boundary conditions are set with $u(0, t) = 0$ and $u(1, t) = 1$. As $t \to \infty$, what do you expect u_n to be for $n = 0, \ldots, 10$? Give an intuitive reason why this should be the case.

16. The Crank–Nicholson method, which has the form

$$u_j^{n+1} = u_j^n + \frac{D\Delta t}{2(\Delta x)^2}(u_{j+1}^n - 2u_j^n + u_{j-1}^n)$$
$$+ \frac{D\Delta t}{2(\Delta x)^2}(u_{j+1}^{n+1} - 2u_j^{n+1} + u_{j-1}^{n+1}),$$

is a well-known implicit method for solving the diffusion equation. Show that its amplification factor g satisfies

$$g = \frac{1 - 2\beta \sin^2(k\Delta x/2)}{1 + 2\beta \sin^2(k\Delta x/2)},$$

where $\beta = D\Delta t/(\Delta x)^2$. Accordingly, show that it is always stable.

References

Ablowitz, M. J., & Segur, H. 1981. *Solitons and the inverse scattering transform*. Vol. 4. Philadelphia: Society for Industrial and Applied Mathematics.

Aki, K., & Richards, P. G. 2002. *Quantitative seismology*. 2nd ed. Sausalito, CA: University Science Books.

Arfken, G. B., & Weber, H.-J. 2005. *Mathematical methods for physicists*. 6th ed. Boston: Elsevier.

Arfken, G. B., Weber, H.-J., & Harris, F. E. 2013. *Mathematical methods for physicists: A comprehensive guide*. 7th ed. Amsterdam: Elsevier.

Ascher, U. M., & Petzold, L. R. 1998. *Computer methods for ordinary differential equations and differential-algebraic equations*. Philadelphia: Society for Industrial and Applied Mathematics.

Backus, G., & Gilbert, F. 1968. The resolving power of gross earth data. *Geophysical Journal International*, **16**(2), 169–205.

Backus, G., & Gilbert, F. 1970. Uniqueness in the inversion of inaccurate gross earth data. *Philosophical Transactions of the Royal Society of London A: Mathematical, Physical and Engineering Sciences*, **266**(1173), 123–192.

Bak, P. 1996. *How nature works: The science of self-organized criticality*. New York: Copernicus.

Barenblatt, G. I. 1996. *Scaling, self-similarity, and intermediate asymptotics*. Vol. 14. Cambridge: Cambridge University Press.

Barenblatt, G. I., & Zel'Dovich, Y. B. 1972. Self-similar solutions as intermediate asymptotics. *Annual Review of Fluid Mechanics*, **4**(1), 285–312.

Batchelor, G. K. 1953. *The theory of homogeneous turbulence*. Cambridge: Cambridge University Press.

Bateman, H., & Erdélyi, A. 1953. *Higher transcendental functions*. New York: McGraw-Hill.

Bender, C. M., & Orszag, S. A. 1999. *Advanced mathematical methods for scientists and engineers*. New York: Springer.

Blackman, R. B., & Tukey, J. W. 1958. *The measurement of power spectra, from the point of view of communications engineering*. New York: Dover Publications.

Boas, M. L. 2006. *Mathematical methods in the physical sciences*. 3rd ed. Hoboken, NJ: Wiley.

Boussinesq, J. 1877. Essai sur la theorie des eaux courantes, Memoires presentes par divers savants. *l'Acad. des Sci. Inst. Nat. France*, **XXIII**, 1–680.

Box, G. E. P., & Muller, M. E. 1958. A note on the generation of random normal deviates. *The Annals of Mathematical Statistics*, **29**(2), 610–611.

Box, G. E. P., Jenkins, G. M., & Reinsel, G. C. 2008. *Time series analysis: Forecasting and control*. 4th ed. Hoboken, NJ: John Wiley.

Bracewell, R. N. 2000. *The Fourier transform and its applications*. 3rd ed. Boston: McGraw Hill.

Brigham, E. O. 1988. *The fast Fourier transform and its applications*. Englewood Cliffs, NJ: Prentice Hall.

Bullard, E. 1955. The stability of a homopolar dynamo. *Mathematical Proceedings of the Cambridge Philosophical Society*, vol. 51, 744–760. Cambridge University Press.

Bullen, K. E. 1963. *An introduction to the theory of seismology*. 3rd ed. Cambridge: Cambridge University Press.

Burgers, J. M. 1948. A mathematical model illustrating the theory of turbulence. *Advances in Applied Mechanics*, **1**, 171–199.

Butkov, E. 1968. *Mathematical physics*. Addison-Wesley series in advanced physics. Reading, MA: Addison-Wesley.

Chandrasekhar, S. 1943. Stochastic problems in physics and astronomy. *Rev. Mod. Phys.*, **15**(Jan), 1–89.

Cheney, E. W. 1982. *Introduction to approximation theory*. 2nd ed. New York: Chelsea.

Churchill, R. V. 1960. *Complex variables and applications*. 2nd ed. New York: McGraw-Hill.

Coddington, E. A., & Levinson, N. 1984. *Theory of ordinary differential equations*. Malabar, FL: R. E. Krieger.

Cody, W. J. 1981. Analysis of proposals for the floating-point standard. *Computer*, 63–68.

Cole, J. D., et al. 1951. On a quasi-linear parabolic equation occurring in aerodynamics. *Quart. Appl. Math*, **9**(3), 225–236.

Cook, A. H. 1973. *Physics of the earth and planets*. New York: Wiley.

Cooley, J. W., & Tukey, J. W. 1965. An algorithm for the machine calculation of complex Fourier series. *Math. Comp.*, **19**(90), 297–301.

Cooley, J. W., Lewis, P. A. W., & Welch, P. D. 1969. The fast Fourier transform and its applications. *IEEE Trans. Education*, **12**(1), 27–34.

Courant, R., & Hilbert, D. 1962. *Methods of mathematical physics*. Wiley classics library. New York: Interscience Publishers.

D'Agostino, R. B., & Stephens, M. A. 1986. *Goodness-of-fit techniques*. Statistics, textbooks and monographs, vol. 68. New York: M. Dekker.

Danby, J. M. A. 1988. *Fundamentals of celestial mechanics*. 2nd ed., rev. & enl. Richmond, VA: Willmann-Bell.

Davis, H. T. 1960. *Introduction to nonlinear differential and integral equations*. U.S. Atomic Energy Commission.

De Pater, I., & Lissauer, J. J. 2010. *Planetary sciences*. 2nd ed. New York: Cambridge University Press.

Deans, S. R. 2007. *The radon transform and some of its applications*. Mineola, NY: Dover Publications.

Drazin, P. G. 1992. *Nonlinear systems*. Cambridge: Cambridge University Press.

Drazin, P. G., & Johnson, R. S. 1989. *Solitons: An introduction*. Cambridge: Cambridge University Press.

Feder, J. 1988. *Fractals*. New York: Plenum Press.

Feigenbaum, M. J. 1980. Universal behavior in nonlinear systems. *Los Alamos Science*, **1**, 4–27.

Feller, W. 1968. *An introduction to probability theory and its applications*. 3rd ed. Vol. 1. New York: John Wiley & Sons.

Feynman, R. P., Leighton, R. B., & Sands, M. L. 1989. *The Feynman lectures on physics*. Redwood City, CA: Addison-Wesley.

Fowler, A. C. 2011. *Mathematical geoscience*. Interdisciplinary applied mathematics, vol. 36. London: Springer.

Freedman, D., Pisani, R., & Purves, R. 2007. *Statistics*. 4th ed. New York: W.W. Norton & Co.

Fröberg, C. E., & Fröberg, C. E. 1985. *Numerical mathematics: Theory and computer applications*. Menlo Park, CA: Benjamin/Cummings.

Garland, G. D. 1979. *Introduction to geophysics: Mantle, core, and crust*. 2nd ed. Philadelphia: Saunders.

Gear, C. W. 1971. *Numerical initial value problems in ordinary differential equations*. Prentice-Hall series in automatic computation. Englewood Cliffs, NJ: Prentice-Hall.

Gentleman, W. M. 1968. Matrix multiplication and fast Fourier transform. *Bell. Syste. Tech. J.*, **47**, 1099–1103.

Glatzmaier, G. A. 2013. *Introduction to modeling convection in planets and stars: Magnetic field, density stratification, rotation*. Princeton series in astrophysics. Princeton, NJ: Princeton University Press.

Gleick, J. 2008. *Chaos: Making a new science*. 20th anniversary ed. New York: Penguin Books.

Goldstein, H., Poole, C. P., & Safko, J. L. 2002. *Classical mechanics*. 3rd ed. San Francisco: Addison Wesley.

Greenberg, M. D. 1998. *Advanced engineering mathematics*. 2nd ed. Upper Saddle River, NJ: Prentice Hall.

Hairer, E., Nørsett, S. P., & Wanner, G. 2009. *Solving ordinary differential equations I: Nonstiff problems*. 2nd rev. ed. Springer series in computational mathematics, vol. 8. Heidelberg: Springer.

Hairer, E., Lubich, C., & Wanner, G. 2010. *Geometric numerical integration: Structure-preserving algorithms for ordinary differential equations*. 2nd ed. Springer series in computational mathematics, vol. 31. Heidelberg: Springer.

Hamming, R. W. 1991. *The art of probability: For scientists and engineers*. Advanced Book Classics. New York: Perseus Books Group.

Harwit, M. 2006. *Astrophysical concepts*. 4th ed. New York: Springer.

Higham, N. J. 2002. *Accuracy and stability of numerical algorithms*. 2nd ed. Philadelphia: Society for Industrial and Applied Mathematics.

Hilborn, R. C. 2000. *Chaos and nonlinear dynamics: An introduction for scientists and engineers*. 2nd ed. Oxford: Oxford University Press.

Hirota, R. 1971. Exact solution of the Korteweg–de Vries equation for multiple collisions of solitons. *Phys. Rev. Lett.*, **27**(Nov), 1192–1194.

Hopf, E. 1950. The partial differential equation $u_t + uu_x = \mu_{xx}$. *Communications on Pure and Applied Mathematics*, **3**(3), 201–230.

Hoppensteadt, F. C. 1982. *Mathematical methods of population biology*. Vol. 4. Cambridge: Cambridge University Press.

Houghton, J. T. 2002. *The physics of atmospheres*. 3rd ed. Cambridge: Cambridge University Press.

Jackson, J. D. 1999. *Classical electrodynamics*. 3rd ed. New York: Wiley.

Jeffreys, H. 1925. On certain approximate solutions of lineae differential equations of the second order. *Proceedings of the London Mathematical Society*, **2**(1), 428–436.

Jeffreys, H., & Jeffreys, B. S. 1999. *Methods of mathematical physics*. 3rd ed. Cambridge: Cambridge University Press.

Jensen, H. J. 1998. *Self-organized criticality: Emergent complex behavior in physical and biological systems*. Vol. 10. Cambridge: Cambridge University Press.

Kaula, W. M. 1968. *An introduction to planetary physics: The terrestrial planets*. Space science text series. New York: Wiley.

Kemble, E. C. 2005. *The fundamental principles of quantum mechanics: With elementary applications*. Mineola, NY: Dover Publications.

Kincaid, D., & Cheney, E. W. 2009. *Numerical analysis: Mathematics of scientific computing*. 3rd ed. The Sally series, vol. 2. Providence, RI: American Mathematical Society.

Knuth, D. E. 1997a. *The art of computer programming: Fundamental algorithms*. Addison-Wesley series in computer science and information processing, vol. 1. Reading, MA: Addison-Wesley.

Knuth, D. E. 1997b. *The art of computer programming: Seminumerical algorithms*. Addison-Wesley series in computer science and information processing, vol. 2. Reading, MA: Addison-Wesley.

Kusse, B., & Westwig, E. 2006. *Mathematical physics: Applied mathematics for scientists and engineers*. 2nd ed. Weinheim: Wiley-VCH.

LeVeque, R. J. 2007. *Finite difference methods for ordinary and partial differential equations: Steady-state and time-dependent problems*. Philadelphia: Society for Industrial and Applied Mathematics.

Lichtenberg, A. J., & Lieberman, M. A. 1992. *Regular and chaotic dynamics*. 2nd ed. Applied mathematical sciences, vol. 38. New York: Springer-Verlag.

Lorenz, E. N. 1963a. Deterministic nonperiodic flow. *Journal of the Atmospheric Sciences*, **20**(2), 130–141.

Lorenz, E. N. 1963b. The predictability of hydrodynamic flow. *Trans. New York Acad of Sciences*, Ser. I, **25**(4), 409–432.

Malhotra, R., Fox, K., Murray, C. D., & Nicholson, P. D. 1989. Secular perturbations of the Uranian satellites - Theory and practice. *Astron. Astrophys.*, **221**(Sept.), 348–358.

Mandelbrot, B. B. 1983. *The fractal geometry of nature*. Updated and augm edn. New York: W. H. Freeman.

Mathews, J., & Walker, R. L. 1970. *Mathematical methods of physics*. 2nd ed. New York: W. A. Benjamin.

May, R. M. 1976. Simple mathematical models with very complicated dynamics. *Nature*, **261**(5560), 459–467.

McKenzie, D. P. 1987. The compaction of igneous and sedimentary rocks. *Journal of the Geological Society*, **144**(2), 299–307.

Monin, A. S., Yaglom, A. M., & Lumley, J. L. 2007. *Statistical fluid mechanics: Mechanics of turbulence*. Mineola, NY: Dover Publications.

Moon, F. C. 1992. *Chaotic and fractal dynamics: An introduction for applied scientists and engineers*. New York: Wiley.

Moran, P. A. P. 1968. *An introduction to probability theory*. Oxford science publications. Clarendon Press.

Morse, P. M., & Feshbach, H. 1999. *Methods of theoretical physics*. Boston: McGraw-Hill.

Murray, C. D., & Dermott, S. F. 1999. *Solar system dynamics*. Cambridge: Cambridge University Press.

Murray, J. D. 2003. *Mathematical biology*. 3rd ed. New York: Springer.

Nayfeh, A. H., & Balachandran, B. 1995. *Applied nonlinear dynamics: Analytical, computational, and experimental methods*. New York: Wiley.

Newman, W. I. 1979a. Analysis of galaxy neutral hydrogen spectra. *Astron. Astrophys.*, **73**(Mar.), 37–39.

Newman, W. I. 1979b. The application of generalized inverse theory to the recovery of temperature profiles. *Journal of the Atmospheric Sciences*, **36**(4), 559–565.

Newman, W. I. 1980. Some exact solutions to a non-linear diffusion problem in population genetics and combustion. *Journal of Theoretical Biology*, **85**(2), 325–334.

Newman, W. I. 1983a. The long-time behavior of the solution to a nonlinear diffusion problem in population genetics and combustion. *Journal of Theoretical Biology*, **104**(4), 473–484.

Newman, W. I. 1983b. Nonlinear diffusion: Self-similarity and traveling-waves. *Pure and Applied Geophysics*, **121**(3), 417–441.

Newman, W. I. 1984. A Lyapunov functional for the evolution of solutions to the porous medium equation to self-similarity. I. *Journal of Mathematical Physics*, **25**(Oct.), 3120–3123.

Newman, W. I. 2000. Inverse cascade via Burgers equation. *Chaos: An Interdisciplinary Journal of Nonlinear Science*, **10**(2), 393–397.

Newman, W. I. 2012. *Continuum mechanics in the earth sciences*. Cambridge: Cambridge University Press.

Newman, W. I., & Efroimsky, M. 2003. The method of variation of constants and multiple time scales in orbital mechanics. *Chaos*, **13**(June), 476–485.

Newman, W. I., & Thorson, W. R. 1972a. New method for rapid numerical solution of the one-dimensional Schrödinger equation. *Physical Review Letters*, **29**(Nov.), 1350–1353.

Newman, W. I., & Thorson, W. R. 1972b. Rapid numerical solution of the one-dimensional Schrödinger equation. *Canadian Journal of Physics*, **50**, 2997.

Nussbaumer, H. J. 1982. *Fast Fourier transform and convolution algorithms*. 2nd corr. and updated ed. Vol. 2. Berlin: Springer-Verlag.

Olver, F. W. J. 2010. *NIST handbook of mathematical functions*. Cambridge: Cambridge University Press.

Parker, R. L. 1977. Understanding inverse theory. *Annual Review of Earth and Planetary Sciences*, **5**, 35.

Parker, R. L. 1994. *Geophysical inverse theory*. Princeton, NJ: Princeton University Press.

Peitgen, H.-O., Saupe, D., & Barnsley, M. F. 1988. *The science of fractal images*. New York: Springer-Verlag.

Peyret, R., & Taylor, T. D. 1990. *Computational methods for fluid flow*. Corr. 3rd print ed. New York: Springer-Verlag.

Potter, D. 1973. *Computational physics*. London: J. Wiley.

Press, W. H., Vetterling, W. T., Teukolsky, S. A., & Flannery, B. P. 1988. *Numerical recipes*. Cambridge: Cambridge University Press.

Ralston, A., & Rabinowitz, P. 2001. *A first course in numerical analysis*. 2nd ed. Mineola, NY: Dover Publications.

Richtmyer, R. D., & Morton, K. W. 1994. *Difference methods for initial-value problems*. 2nd ed. Malabar, FL: Krieger.

Roy, A. E. 2005. *Orbital motion*. 4th ed. Bristol, England: Institute of Physics.

Sanz-Serna, J. M., & Calvo, M. P. 1994. *Numerical Hamiltonian problems*. 1st ed. Vol. 7. London: Chapman & Hall.

Schiesser, W. E. 1991. *The numerical method of lines: Integration of partial differential equations*. San Diego: Academic Press.

Scott, D. R., & Stevenson, D. J. 1984. Magma solitons. *Geophys. Res. Lett.*, **11**, 1161–1164.

Scott, D. R., & Stevenson, D. J. 1986. Magma ascent by porous flow. *J. Geophys. Res.*, **91**(Aug.), 9283–9296.

Scott, D. R., Stevenson, D. J., & Whitehead, J. A. 1986. Observations of solitary waves in a viscously deformable pipe. *Nature*, **319**(Feb.), 759–761.

Shearer, P. M. 2009. *Introduction to seismology*. 2nd ed. Cambridge: Cambridge University Press.

Silver, N. 2012. *The signal and the noise: Why so many predictions fail but some don't*. New York: Penguin Press.

Sneddon, I. N. 1972. *The use of integral transforms*. New York: McGraw-Hill.

Sornette, D. 2006. *Critical phenomena in natural sciences: Chaos, fractals, selforganization, and disorder: Concepts and tools*. 2nd ed. Springer series in synergetics. Berlin: Springer.

Sparrow, C. 2005. *The Lorenz equations: Bifurcations, chaos, and strange attractors*. Mineola, NY: Dover Publications.

Stacey, F. D., & Davis, P. M. 2008. *Physics of the Earth*. 4th ed. Cambridge: Cambridge University Press.

Stakgold, I. 1998. *Green's functions and boundary value problems*. 2nd ed. Pure and applied mathematics. New York: Wiley.

Stone, M., & Goldbart, P. M. 2009. *Mathematics for physics: A guided tour for graduate students*. Cambridge: Cambridge University Press.

Strang, G. 1986. *Introduction to applied mathematics*. Wellesley, MA: Wellesley-Cambridge Press.

Strogatz, S. H. 1994. *Nonlinear dynamics and chaos: With applications to physics, biology, chemistry, and engineering*. Reading, MA: Addison-Wesley.

Taff, L. G. 1985. *Celestial mechanics: A computational guide for the practitioner*. New York: Wiley.

Tennekes, H., & Lumley, J. L. 1972. *A first course in turbulence*. Cambridge, MA: MIT Press.

Turcotte, D. L. 1997. *Fractals and chaos in geology and geophysics*. 2nd ed. Cambridge: Cambridge University Press.

Turcotte, D. L., & Newman, W. I. 1996. Symmetries in geology and geophysics. *Proceedings of the National Academy of Sciences*, **93**(25), 14295–14300.

Turcotte, D. L., Schubert, G., & Turcotte, D. L. 2002. *Geodynamics*. 2nd ed. Cambridge: Cambridge University Press.

Walker, J. S. 1988. *Fourier analysis*. New York: Oxford University Press.

Watson, G. N. 1995. *A treatise on the theory of Bessel functions*. 2nd ed., Cambridge mathematical library ed. Cambridge: Cambridge University Press.

Weber, H.-J., & Arfken, G. B. 2004. *Essential mathematical methods for physicists*. San Diego, CA: Academic Press.

Whitham, G. B. 1974. *Linear and nonlinear waves*. Pure and applied mathematics. New York: Wiley.

Whitney, C. A. 1990. *Random processes in physical systems: An introduction to probability-based computer simulations*. A Wiley-Interscience publication. New York: Wiley.

Whittaker, E. T., & Watson, G. N. 1979. *A course of modern analysis: An introduction to the general theory of infinite processes and of analytic functions, with an account of the principal transcendental functions*. 1st AMS ed. New York: AMS Press.

Wilkinson, J. H. 1959. The evaluation of the zeros of ill-conditioned polynomials. Part I. *Numerische Mathematik*, **1**, 150–166.

Index